An Introduction to Econophysics

This book concerns the use of concepts from statistical physics in the description of financial systems. Specifically, the authors illustrate the scaling concepts used in probability theory, in critical phenomena, and in fully developed turbulent fluids. These concepts are then applied to financial time series to gain new insights into the behavior of financial markets. The authors also present a new stochastic model that displays several of the statistical properties observed in empirical data.

Usually in the study of economic systems it is possible to investigate the system at different scales. But it is often impossible to write down the 'microscopic' equation for all the economic entities interacting within a given system. Statistical physics concepts such as stochastic dynamics, short- and long-range correlations, self-similarity and scaling permit an understanding of the global behavior of economic systems without first having to work out a detailed microscopic description of the same system. This book will be of interest both to physicists and to economists. Physicists will find the application of statistical physics concepts to economic systems interesting and challenging, as economic systems are among the most intriguing and fascinating complex systems that might be investigated. Economists and workers in the financial world will find useful the presentation of empirical analysis methods and well-formulated theoretical tools that might help describe systems composed of a huge number of interacting subsystems.

This book is intended for students and researchers studying economics or physics at a graduate level and for professionals in the field of finance. Undergraduate students possessing some familarity with probability theory or statistical physics should also be able to learn from the book.

DR ROSARIO N. MANTEGNA is interested in the empirical and theoretical modeling of complex systems. Since 1989, a major focus of his research has been studying financial systems using methods of statistical physics. In particular, he has originated the theoretical model of the truncated Lévy flight and discovered that this process describes several of the statistical properties of the Standard and Poor's 500 stock index. He has also applied concepts of ultrametric spaces and cross-correlations to the modeling of financial markets. Dr Mantegna is a Professor of Physics at the University of Palermo.

DR H. EUGENE STANLEY has served for 30 years on the physics faculties of MIT and Boston University. He is the author of the 1971 monograph *Introduction to Phase Transitions and Critical Phenomena* (Oxford University Press, 1971). This book brought to a much wider audience the key ideas of scale invariance that have proved so useful in various fields of scientific endeavor. Recently, Dr Stanley and his collaborators have been exploring the degree to which scaling concepts give insight into economics and various problems of relevance to biology and medicine.

AN INTRODUCTION TO ECONOPHYSICS
Correlations and Complexity in Finance

ROSARIO N. MANTEGNA

Dipartimento di Energetica ed Applicazioni di Fisica, Palermo University

H. EUGENE STANLEY

Center for Polymer Studies and Department of Physics, Boston University

CAMBRIDGE
UNIVERSITY PRESS

CAMBRIDGE UNIVERSITY PRESS
Cambridge, New York, Melbourne, Madrid, Cape Town, Singapore, São Paulo

Cambridge University Press
The Edinburgh Building, Cambridge CB2 8RU, UK

Published in the United States of America by Cambridge University Press, New York

www.cambridge.org
Information on this title: www.cambridge.org/9780521620086

First published 2000
Fourth printing 2004
This digitally printed version 2007

A catalogue record for this publication is available from the British Library

Library of Congress Cataloguing in Publication data
Mantegna, Rosario N. (Rosario Nunzio), 1960–
An introduction to econophysics: correlations and complexity in
finance / Rosario N. Mantegna, H. Eugene Stanley.
p. cm.
ISBN 0 521 62008 2 (hardbound)
1. Finance–Statistical methods. 2. Finance–Mathematical models.
3. Statistical physics. I. Stanley, H. Eugene (Harry Eugene),
1941– . II. Title
HG176.5.M365 1999
332'.01'5195–dc21 99-28047 CIP

ISBN 978-0-521-62008-6 hardback
ISBN 978-0-521-03987-1 paperback

Contents

Preface

Physicists are currently contributing to the modeling of 'complex systems' by using tools and methodologies developed in statistical mechanics and theoretical physics. Financial markets are remarkably well-defined complex systems, which are continuously monitored – down to time scales of seconds. Further, virtually every economic transaction is recorded, and an increasing fraction of the total number of recorded economic data is becoming accessible to interested researchers. Facts such as these make financial markets extremely attractive for researchers interested in developing a deeper understanding of modeling of complex systems.

Economists – and mathematicians – are the researchers with the longer tradition in the investigation of financial systems. Physicists, on the other hand, have generally investigated economic systems and problems only occasionally. Recently, however, a growing number of physicists is becoming involved in the analysis of economic systems. Correspondingly, a significant number of papers of relevance to economics is now being published in physics journals. Moreover, new interdisciplinary journals – and dedicated sections of existing journals – have been launched, and international conferences are being organized.

In addition to fundamental issues, practical concerns may explain part of the recent interest of physicists in finance. For example, risk management, a key activity in financial institutions, is a complex task that benefits from a multidisciplinary approach. Often the approaches taken by physicists are complementary to those of more established disciplines, so including physicists in a multidisciplinary risk management team may give a cutting edge to the team, and enable it to succeed in the most efficient way in a competitive environment.

This book is designed to introduce the multidisciplinary field of econophysics, a neologism that denotes the activities of physicists who are working on economics problems to test a variety of new conceptual approaches deriving from the physical sciences. The book is short, and is not designed to review all the recent work done in this rapidly developing area. Rather, the book offers an introduction that is sufficient to allow the current literature to be profitably read. Since this literature spans disciplines ranging from financial mathematics and probability theory to physics and economics, unavoidable notation confusion is minimized by including a systematic notation list in the appendix.

We wish to thank many colleagues for their assistance in helping prepare this book. Various drafts were kindly criticized by Andreas Buchleitner, Giovanni Bonanno, Parameswaran Gopikrishnan, Fabrizio Lillo, Johannes Voigt, Dietrich Stauffer, Angelo Vulpiani, and Dietrich Wolf.

Jerry D. Morrow demonstrated his considerable T$_E$X skills in carrying out the countless revisions required. Robert Tomposki's tireless library research greatly improved the bibliography. We especially thank the staff of Cambridge University Press – most especially Simon Capelin (Publishing Director in the Physical Sciences), Sue Tuck (Production Controller), and Lindsay Nightingale (Copy Editor), and the CUP Technical Applications Group – for their remarkable efficiency and good cheer throughout this entire project.

As we study the final page proof, we must resist the strong urge to re-write the treatment of several topics that we now realize can be explained more clearly and precisely. We do hope that readers who notice these and other imperfections will communicate their thoughts to us.

Rosario N. Mantegna

H. Eugene Stanley

To Francesca and Idahlia

1

Introduction

1.1 Motivation

Since the 1970s, a series of significant changes has taken place in the world of finance. One key year was 1973, when currencies began to be traded in financial markets and their values determined by the foreign exchange market, a financial market active 24 hours a day all over the world. During that same year, Black and Scholes [18] published the first paper that presented a rational option-pricing formula.

Since that time, the volume of foreign exchange trading has been growing at an impressive rate. The transaction volume in 1995 was 80 times what it was in 1973. An even more impressive growth has taken place in the field of derivative products. The total value of financial derivative market contracts issued in 1996 was 35 trillion US dollars. Contracts totaling approximately 25 trillion USD were negotiated in the over-the-counter market (i.e., directly between firms or financial institutions), and the rest (approximately 10 trillion USD) in specialized exchanges that deal only in derivative contracts. Today, financial markets facilitate the trading of huge amounts of money, assets, and goods in a competitive global environment.

A second revolution began in the 1980s when electronic trading, already a part of the environment of the major stock exchanges, was adapted to the foreign exchange market. The electronic storing of data relating to financial contracts – or to prices at which traders are willing to buy (bid quotes) or sell (ask quotes) a financial asset – was put in place at about the same time that electronic trading became widespread. One result is that today a huge amount of electronically stored financial data is readily available. These data are characterized by the property of being high-frequency data – the average time delay between two records can be as short as a few seconds. The enormous expansion of financial markets requires strong investments in money and

1

human resources to achieve reliable quantification and minimization of risk for the financial institutions involved.

1.2 Pioneering approaches

In this book we discuss the application to financial markets of such concepts as power-law distributions, correlations, scaling, unpredictable time series, and random processes. During the past 30 years, physicists have achieved important results in the field of phase transitions, statistical mechanics, nonlinear dynamics, and disordered systems. In these fields, power laws, scaling, and unpredictable (stochastic or deterministic) time series are present and the current interpretation of the underlying physics is often obtained using these concepts.

With this background in mind, it may surprise scholars trained in the natural sciences to learn that the first use of a power-law distribution – and the first mathematical formalization of a random walk – took place in the social sciences. Almost exactly 100 years ago, the Italian social economist Pareto investigated the statistical character of the wealth of individuals in a stable economy by modeling them using the distribution

$$y \sim x^{-v}, \tag{1.1}$$

where y is the number of people having income x or greater than x and v is an exponent that Pareto estimated to be 1.5 [132]. Pareto noticed that his result was quite general and applicable to nations 'as different as those of England, of Ireland, of Germany, of the Italian cities, and even of Peru'.

It should be fully appreciated that the concept of a power-law distribution is counterintuitive, because it may lack any characteristic scale. This property prevented the use of power-law distributions in the natural sciences until the recent emergence of new paradigms (i) in probability theory, thanks to the work of Lévy [92] and thanks to the application of power-law distributions to several problems pursued by Mandelbrot [103]; and (ii) in the study of phase transitions, which introduced the concepts of scaling for thermodynamic functions and correlation functions [147].

Another concept ubiquitous in the natural sciences is the random walk. The first theoretical description of a random walk in the natural sciences was performed in 1905 by Einstein [48] in his famous paper dealing with the determination of the Avogadro number. In subsequent years, the mathematics of the random walk was made more rigorous by Wiener [158], and now the random walk concept has spread across almost all research areas in the natural sciences.

The first formalization of a random walk was not in a publication by Einstein, but in a doctoral thesis by Bachelier [8]. Bachelier, a French mathematician, presented his thesis to the faculty of sciences at the Academy of Paris on 29 March 1900, for the degree of *Docteur en Sciences Mathématiques*. His advisor was Poincaré, one of the greatest mathematicians of his time. The thesis, entitled *Théorie de la spéculation*, is surprising in several respects. It deals with the pricing of options in speculative markets, an activity that today is extremely important in financial markets where derivative securities – those whose value depends on the values of other more basic underlying variables – are regularly traded on many different exchanges. To complete this task, Bachelier determined the probability of price changes by writing down what is now called the Chapman–Kolmogorov equation and recognizing that what is now called a Wiener process satisfies the diffusion equation (this point was rediscovered by Einstein in his 1905 paper on Brownian motion). Retrospectively analyzed, Bachelier's thesis lacks rigor in some of its mathematical and economic points. Specifically, the determination of a Gaussian distribution for the price changes was – mathematically speaking – not sufficiently motivated. On the economic side, Bachelier investigated price changes, whereas economists are mainly dealing with changes in the logarithm of price. However, these limitations do not diminish the value of Bachelier's pioneering work.

To put Bachelier's work into perspective, the Black & Scholes option-pricing model – considered the milestone in option-pricing theory – was published in 1973, almost three-quarters of a century after the publication of his thesis. Moreover, theorists and practitioners are aware that the Black & Scholes model needs correction in its application, meaning that the problem of which stochastic process describes the changes in the logarithm of prices in a financial market is still an open one.

The problem of the distribution of price changes has been considered by several authors since the 1950s, which was the period when mathematicians began to show interest in the modeling of stock market prices. Bachelier's original proposal of Gaussian distributed price changes was soon replaced by a model in which stock prices are log-normal distributed, i.e., stock prices are performing a geometric Brownian motion. In a geometric Brownian motion, the differences of the logarithms of prices are Gaussian distributed. This model is known to provide only a first approximation of what is observed in real data. For this reason, a number of alternative models have been proposed with the aim of explaining

(i) the empirical evidence that the tails of measured distributions are fatter than expected for a geometric Brownian motion; and
(ii) the time fluctuations of the second moment of price changes.

Among the alternative models proposed, 'the most revolutionary development in the theory of speculative prices since Bachelier's initial work' [38], is Mandelbrot's hypothesis that price changes follow a Lévy stable distribution [102]. Lévy stable processes are stochastic processes obeying a generalized central limit theorem. By obeying a generalized form of the central limit theorem, they have a number of interesting properties. They are stable (as are the more common Gaussian processes) – i.e., the sum of two independent stochastic processes x_1 and x_2 characterized by the same Lévy distribution of index α is itself a stochastic process characterized by a Lévy distribution of the same index. The shape of the distribution is maintained (is stable) by summing up independent identically distributed Lévy stable random variables.

As we shall see, Lévy stable processes define a basin of attraction in the functional space of probability density functions. The sum of independent identically distributed stochastic processes $S_n \equiv \sum_{i=1}^{n} x_i$ characterized by a probability density function with power-law tails,

$$P(x) \sim x^{-(1+\alpha)}, \tag{1.2}$$

will converge, in probability, to a Lévy stable stochastic process of index α when n tends to infinity [66].

This property tells us that the distribution of a Lévy stable process is a power-law distribution for large values of the stochastic variable x. The fact that power-law distributions may lack a typical scale is reflected in Lévy stable processes by the property that the variance of Lévy stable processes is infinite for $\alpha < 2$. Stochastic processes with infinite variance, although well defined mathematically, are extremely difficult to use and, moreover, raise fundamental questions when applied to real systems. For example, in physical systems the second moment is often related to the system temperature, so infinite variances imply an infinite (or undefined) temperature. In financial systems, an infinite variance would complicate the important task of risk estimation.

1.3 The chaos approach

A widely accepted belief in financial theory is that time series of asset prices are unpredictable. This belief is the cornerstone of the description of price dynamics as stochastic processes. Since the 1980s it has been recognized in the physical sciences that unpredictable time series and stochastic processes are not synonymous. Specifically, chaos theory has shown that unpredictable time series can arise from deterministic nonlinear systems. The results obtained in the study of physical and biological systems triggered an interest in economic systems, and theoretical and empirical studies have investigated whether the time evolution of asset prices in financial markets might indeed be due to underlying nonlinear deterministic dynamics of a (limited) number of variables.

One of the goals of researchers studying financial markets with the tools of nonlinear dynamics has been to reconstruct the (hypothetical) strange attractor present in the chaotic time evolution and to measure its dimension d. The reconstruction of the underlying attractor and its dimension d is not an easy task. The more reliable estimation of d is the inequality $d > 6$. For chaotic systems with $d > 3$, it is rather difficult to distinguish between a chaotic time evolution and a random process, especially if the underlying deterministic dynamics are unknown. Hence, from an empirical point of view, it is quite unlikely that it will be possible to discriminate between the random and the chaotic hypotheses.

Although it cannot be ruled out that financial markets follow chaotic dynamics, we choose to work within a paradigm that asserts price dynamics are stochastic processes. Our choice is motivated by the observation that the time evolution of an asset price depends on all the information affecting (or believed to be affecting) the investigated asset and it seems unlikely to us that all this information can be essentially described by a small number of nonlinear deterministic equations.

1.4 The present focus

Financial markets exhibit several of the properties that characterize complex systems. They are open systems in which many subunits interact nonlinearly in the presence of feedback. In financial markets, the governing rules are rather stable and the time evolution of the system is continuously monitored. It is now possible to develop models and to test their accuracy and predictive power using available data, since large databases exist even for high-frequency data.

One of the more active areas in finance is the pricing of derivative instruments. In the simplest case, an asset is described by a stochastic process and a derivative security (or contingent claim) is evaluated on the basis of the type of security and the value and statistical properties of the underlying asset. This problem presents at least two different aspects: (i) 'fundamental' aspects, which are related to the nature of the random process of the asset, and (ii) 'applied' or 'technical' aspects, which are related to the solution of the option-pricing problem under the assumption that the underlying asset performs the proposed random process.

Recently, a growing number of physicists have attempted to analyze and model financial markets and, more generally, economic systems. The interest of this community in financial and economic systems has roots that date back to 1936, when Majorana wrote a pioneering paper on the essential analogy between statistical laws in physics and in the social sciences [101]. This unorthodox point of view was considered of marginal interest until recently. Indeed, prior to the 1990s, very few professional physicists did any research associated with social or economic systems. The exceptions included Kadanoff [76], Montroll [125], and a group of physical scientists at the Santa Fe Institute [5].

Since 1990, the physics research activity in this field has become less episodic and a research community has begun to emerge. New interdisci-plinary journals have been published, conferences have been organized, and a set of potentially tractable scientific problems has been provisionally iden-tified. The research activity of this group of physicists is complementary to the most traditional approaches of finance and mathematical finance. One characteristic difference is the emphasis that physicists put on the empir-ical analysis of economic data. Another is the background of theory and method in the field of statistical physics developed over the past 30 years that physicists bring to the subject. The concepts of scaling, universality, disordered frustrated systems, and self-organized systems might be helpful in the analysis and modeling of financial and economic systems. One argument that is sometimes raised at this point is that an empirical analysis performed on financial or economic data is not equivalent to the usual experimental investigation that takes place in physical sciences. In other words, it is im-possible to perform large-scale experiments in economics and finance that could falsify any given theory.

We note that this limitation is not specific to economic and financial systems, but also affects such well developed areas of physics as astrophysics, atmospheric physics, and geophysics. Hence, in analogy to activity in these more established areas, we find that we are able to test and falsify any theories associated with the currently available sets of financial and economic data provided in the form of recorded files of financial and economic activity.

Among the important areas of physics research dealing with financial and economic systems, one concerns the complete statistical characterization of the stochastic process of price changes of a financial asset. Several studies have been performed that focus on different aspects of the analyzed stochastic process, e.g., the shape of the distribution of price changes [22, 64, 67, 105, 111, 135], the temporal memory [35, 93, 95, 112], and the higher-order statistical properties [6, 31, 126]. This is still an active area, and attempts are ongoing to develop the most satisfactory stochastic model describing all the features encountered in empirical analyses. One important accomplishment in this area is an almost complete consensus concerning the finiteness of the second moment of price changes. This has been a longstanding problem in finance, and its resolution has come about because of the renewed interest in the empirical study of financial systems.

A second area concerns the development of a theoretical model that is able to encompass all the essential features of real financial markets. Several models have been proposed [10, 11, 23, 25, 29, 90, 91, 104, 117, 142, 146, 149–152], and some of the main properties of the stochastic dynamics of stock price are reproduced by these models as, for example, the leptokurtic 'fat-tailed' non-Gaussian shape of the distribution of price differences. Parallel attempts in the modeling of financial markets have been developed by economists [98–100].

Other areas that are undergoing intense investigations deal with the rational pricing of a derivative product when some of the canonical assumptions of the Black & Scholes model are relaxed [7, 21, 22] and with aspects of portfolio selection and its dynamical optimization [14, 62, 63, 116, 145]. A further area of research considers analogies and differences between price dynamics in a financial market and such physical processes as turbulence [64, 112, 113] and ecological systems [55, 135].

One common theme encountered in these research areas is the time correlation of a financial series. The detection of the presence of a higher-order correlation in price changes has motivated a reconsideration of some beliefs of what is termed 'technical analysis' [155].

In addition to the studies that analyze and model financial systems, there are studies of the income distribution of firms and studies of the statistical properties of their growth rates [2, 3, 148, 153]. The statistical properties of the economic performances of complex organizations such as universities or entire countries have also been investigated [89].

This brief presentation of some of the current efforts in this emerging discipline has only illustrative purposes and cannot be exhaustive. For a more complete overview, consider, for example, the proceedings of conferences dedicated to these topics [78, 88, 109].

2
Efficient market hypothesis

2.1 Concepts, paradigms, and variables

Financial markets are systems in which a large number of traders interact
with one another and react to external information in order to determine
the best price for a given item. The goods might be as different as animals,
ore, equities, currencies, or bonds – or derivative products issued on those
underlying financial goods. Some markets are localized in specific cities (e.g.,
New York, Tokyo, and London) while others (such as the foreign exchange
market) are delocalized and accessible all over the world.

When one inspects a time series of the time evolution of the price, volume,
and number of transactions of a financial product, one recognizes that the
time evolution is unpredictable. At first sight, one might sense a curious
paradox. An important time series, such as the price of a financial good,
is essentially indistinguishable from a stochastic process. There are deep
reasons for this kind of behavior, and in this chapter we will examine some
of these.

2.2 Arbitrage

A key concept for the understanding of markets is the concept of arbitrage
– the purchase and sale of the same or equivalent security in order to profit
from price discrepancies. Two simple examples illustrate this concept. At a
given time, 1 kg of oranges costs 0.60 euro in Naples and 0.50 USD in
Miami. If the cost of transporting and storing 1 kg of oranges from Miami
to Naples is 0.10 euro, by buying 100,000 kg of oranges in Miami and
immediately selling them in Naples it is possible to realize a risk-free profit
of

$$100,000[0.60 - (0.80 \times 0.50) - 0.10] = 10,000 \text{ euro.} \tag{2.1}$$

Here it is assumed that the exchange rate between the US dollar and the euro is 0.80 at the time of the transaction.

This kind of arbitrage opportunity can also be observed in financial markets. Consider the following situation. A stock is traded in two different stock exchanges in two countries with different currencies, e.g., Milan and New York. The current price of a share of the stock is 9 USD in New York and 8 euro in Milan and the exchange rate between USD and euro is 0.80. By buying 1,000 shares of the stock in New York and selling them in Milan, the arbitrager makes a profit (apart from transaction costs) of

$$1,000 \left(\frac{8}{0.80} - 9 \right) = 1,000 \text{ USD}. \tag{2.2}$$

The presence of traders looking for arbitrage conditions contributes to a market's ability to evolve the most rational price for a good. To see this, suppose that one has discovered an arbitrage opportunity. One will exploit it and, if one succeeds in making a profit, one will repeat the same action. In the above example, oranges are bought in Miami and sold in Naples. If this action is carried out repeatedly and systematically, the demand for oranges will increase in Miami and decrease in Naples. The net effect of this action will then be an increase in the price of oranges in Miami and a decrease in the price in Naples. After a period of time, the prices in both locations will become more 'rational', and thus will no longer provide arbitrage opportunities.

To summarize: (i) new arbitrage opportunities continually appear and are discovered in the markets but (ii) as soon as an arbitrage opportunity begins to be exploited, the system moves in a direction that gradually eliminates the arbitrage opportunity.

2.3 Efficient market hypothesis

Markets are complex systems that incorporate information about a given asset in the time series of its price. The most accepted paradigm among scholars in finance is that the market is highly efficient in the determination of the most rational price of the traded asset. The efficient market hypothesis was originally formulated in the 1960s [53]. A market is said to be efficient if all the available information is instantly processed when it reaches the market and it is immediately reflected in a new value of prices of the assets traded.

The theoretical motivation for the efficient market hypothesis has its roots in the pioneering work of Bachelier [8], who at the beginning of the twentieth century proposed that the price of assets in a speculative market be described as a stochastic process. This work remained almost unknown until the 1950s, when empirical results [38] about the serial correlation of the rate of return showed that correlations on a short time scale are negligible and that the approximate behavior of return time series is indeed similar to uncorrelated random walks.

The efficient market hypothesis was formulated explicitly in 1965 by Samuelson [141], who showed mathematically that properly anticipated prices fluctuate randomly. Using the hypothesis of rational behavior and market efficiency, he was able to demonstrate how Y_{t+1}, the expected value of the price of a given asset at time $t+1$, is related to the previous values of prices $Y_0, Y_1, \ldots, Y_t$ through the relation

$$E\{Y_{t+1}|Y_0, Y_1, \ldots, Y_t\} = Y_t. \tag{2.3}$$

Stochastic processes obeying the conditional probability given in Eq. (2.3) are called martingales (see Appendix B for a formal definition). The notion of a martingale is, intuitively, a probabilistic model of a 'fair' game. In gambler's terms, the game is fair when gains and losses cancel, and the gambler's expected future wealth coincides with the gambler's present assets. The fair game conclusion about the price changes observed in a financial market is equivalent to the statement that there is no way of making a profit on an asset by simply using the recorded history of its price fluctuations. The conclusion of this 'weak form' of the efficient market hypothesis is then that price changes are unpredictable from the historical time series of those changes.

Since the 1960s, a great number of empirical investigations have been devoted to testing the efficient market hypothesis [54]. In the great majority of the empirical studies, the time correlation between price changes has been found to be negligibly small, supporting the efficient market hypothesis. However, it was shown in the 1980s that by using the information present in additional time series such as earnings/price ratios, dividend yields, and term-structure variables, it is possible to make predictions of the rate of return of a given asset on a long time scale, much longer than a month. Thus empirical observations have challenged the stricter form of the efficient market hypothesis.

Thus empirical observations and theoretical considerations show that price changes are difficult if not impossible to predict if one starts from the time series of price changes. In its strict form, an efficient market is an idealized system. In actual markets, residual inefficiencies are always present. Searching out and exploiting arbitrage opportunities is one way of eliminating market inefficiencies.

2.4 Algorithmic complexity theory

The description of a fair game in terms of a martingale is rather formal. In this section we will provide an explanation – in terms of information theory and algorithmic complexity theory – of why the time series of returns appears to be random. Algorithmic complexity theory was developed independently by Kolmogorov [85] and Chaitin [28] in the mid-1960s, by chance during the same period as the application of the martingale to economics.

Within algorithmic complexity theory, the complexity of a given object coded in an n-digit binary sequence is given by the bit length $K^{(n)}$ of the shortest computer program that can print the given symbolic sequence. Kolmogorov showed that such an algorithm exists; he called this algorithm asymptotically optimal.

To illustrate this concept, suppose that as a part of space exploration we want to transport information about the scientific and social achievements of the human race to regions outside the solar system. Among the information blocks we include, we transmit the value of π expressed as a decimal carried out to 125,000 places and the time series of the daily values of the Dow–Jones industrial average between 1898 and the year of the space exploration (approximately 125,000 digits). To minimize the amount of storage space and transmission time needed for these two items of information, we write the two number sequences using, for each series, an algorithm that makes use of the regularities present in the sequence of digits. The best algorithm found for the sequence of digits in the value of π is extremely short. In contrast, an algorithm with comparable efficiency has not been found for the time series of the Dow–Jones index. The Dow–Jones index time series is a nonredundant time series.

Within algorithmic complexity theory, a series of symbols is considered unpredictable if the information embodied in it cannot be 'compressed' or reduced to a more compact form. This statement is made more formal by saying that the most efficient algorithm reproducing the original series of symbols has the same length as the symbol sequence itself.

Algorithmic complexity theory helps us understand the behavior of a financial time series. In particular:

(i) Algorithmic complexity theory makes a clearer connection between the efficient market hypothesis and the unpredictable character of stock returns. Such a connection is now supported by the property that a time series that has a dense amount of nonredundant economic information (as the efficient market hypothesis requires for the stock returns time series) exhibits statistical features that are almost indistinguishable from those observed in a time series that is random.

(ii) Measurements of the deviation from randomness provide a tool to verify the validity and limitations of the efficient market hypothesis.

(iii) From the point of view of algorithmic complexity theory, it is impossible to discriminate between trading on 'noise' and trading on 'information' (where now we use 'information' to refer to fundamental information concerning the traded asset, internal or external to the market). Algorithmic complexity theory detects no difference between a time series carrying a large amount of nonredundant economic information and a pure random process.

2.5 Amount of information in a financial time series

Financial time series look unpredictable, and their future values are essentially impossible to predict. This property of the financial time series is not a manifestation of the fact that the time series of price of financial assets does not reflect any valuable and important economic information. Indeed, the opposite is true. The time series of the prices in a financial market carries a large amount of nonredundant information. Because the quantity of this information is so large, it is difficult to extract a subset of economic information associated with some specific aspect. The difficulty in making predictions is thus related to an abundance of information in the financial data, not to a lack of it. When a given piece of information affects the price in a market in a specific way, the market is not completely efficient. This allows us to detect, from the time series of price, the presence of this information. In similar cases, arbitrage strategies can be devised and they will last until the market recovers efficiency in mixing all the sources of information during the price formation.

2.6 Idealized systems in physics and finance

The efficient market is an idealized system. Real markets are only approximately efficient. This fact will probably not sound too unfamiliar to physicists because they are well acquainted with the study of idealized systems. Indeed, the use of idealized systems in scientific investigation has been instrumental in the development of physics as a discipline. Where would physics be without idealizations such as frictionless motion, reversible transformations in thermodynamics, and infinite systems in the critical state? Physicists use these abstractions in order to develop theories and to design experiments. At the same time, physicists always remember that idealized systems only approximate real systems, and that the behavior of real systems will always deviate from that of idealized systems. A similar approach can be taken in the study of financial systems. We can assume realistic 'ideal' conditions, e.g., the existence of a perfectly efficient market, and within this ideal framework develop theories and perform empirical tests. The validity of the results will depend on the validity of the assumptions made.

The concept of the efficient market is useful in any attempt to model financial markets. After accepting this paradigm, an important step is to fully characterize the statistical properties of the random processes observed in financial markets. In the following chapters, we will see that this task is not straightforward, and that several advanced concepts of probability theory are required to achieve a satisfactory description of the statistical properties of financial market data.

3

Random walk

In this chapter we discuss some statistical properties of a random walk. Specifically, (i) we discuss the central limit theorem, (ii) we consider the scaling properties of the probability densities of walk increments, and (iii) we present the concept of asymptotic convergence to an attractor in the functional space of probability densities.

3.1 One-dimensional discrete case

Consider the sum of n independent identically distributed (i.i.d.) random variables x_i,

$$S_n \equiv x_1 + x_2 + \cdots + x_n. \tag{3.1}$$

Here $S_n \equiv x(n\Delta t)$ can be regarded as the sum of n random variables or as the position of a single walker at time $t = n\Delta t$, where n is the number of steps performed, and Δt the time interval required to perform one step. Identically distributed random variables $\{x_i\}$ are characterized by moments $E\{x_i^n\}$ that do not depend on i.

The simplest example is a walk performed by taking random steps of size s, so x_i randomly takes the values $\pm s$. The first and second moments for such a process are

$$E\{x_i\} = 0 \quad \text{and} \quad E\{x_i^2\} = s^2. \tag{3.2}$$

For this random walk

$$E\{x_i x_j\} = \delta_{ij} s^2. \tag{3.3}$$

From (3.1)–(3.3), it follows that

$$E\{x(n\Delta t)\} = \sum_{i=1}^{n} E\{x_i\} = 0, \tag{3.4}$$

14

and

$$E\{x^2(n\Delta t)\} = \sum_{i=1}^{n}\sum_{j=1}^{n} E\{x_i x_j\} = \sum_{i=1}^{n} E\{x_i^2\} = ns^2. \tag{3.5}$$

For a random walk, the variance of the process grows linearly with the number of steps n. Starting from the discrete random walk, a continuous limit can be constructed, as described in the next section.

3.2 The continuous limit

The continuous limit of a random walk may be achieved by considering the limit $n \to \infty$ and $\Delta t \to 0$ such that $t \equiv n\Delta t$ is finite. Then

$$E\{x^2(t)\} = ns^2 = \frac{s^2}{\Delta t} t. \tag{3.6}$$

To have consistency in the limits $n \to \infty$ or $\Delta t \to 0$ with $s^2 = D\Delta t$, it follows that

$$E\{x^2(t)\} = Dt. \tag{3.7}$$

The linear dependence of the variance $x^2(t)$ on t is characteristic of a diffusive process, and D is termed the diffusion constant.

This stochastic process is called a *Wiener process*. Usually it is implicitly assumed that for $n \to \infty$ or $\Delta t \to 0$, the stochastic process $x(t)$ is a Gaussian process. The equivalence

'random walk' $\equiv$ 'Gaussian walk'

holds only when $n \to \infty$ and is not generally true in the discrete case when n is finite, since S_n is characterized by a probability density function (pdf) that is, in general, non-Gaussian and that assumes the Gaussian shape only asymptotically with n. The pdf of the process, $P[x(n\Delta t)]$ – or equivalently $P(S_n)$ – is a function of n, and $P(x_i)$ is arbitrary.

How does the shape of $P[x(n\Delta t)]$ change with time? Under the assumption of independence,

$$P[x(2\Delta t)] = P(x_1) \bigotimes P(x_2), \tag{3.8}$$

where $\bigotimes$ denotes the convolution. In Fig. 3.1 we show four different pdfs $P(x)$: (i) a delta distribution, (ii) a uniform distribution, (iii) a Gaussian distribution, and (iv) a Lorentzian (or Cauchy) distribution. When one of these distributions characterizes the random variables x_i, the pdf $P(S_n)$ changes as n increases (Fig. 3.2).

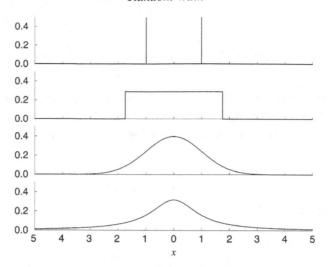

Fig. 3.1. Examples of different probability density functions (pdfs). From top to bottom are shown (i) $P(x) = \delta(x+1)/2 + \delta(x-1)/2$, (ii) a uniform pdf with zero mean and unit standard deviation, (iii) a Gaussian pdf with zero mean and unit standard deviation, and (iv) a Lorentzian pdf with unit scale factor.

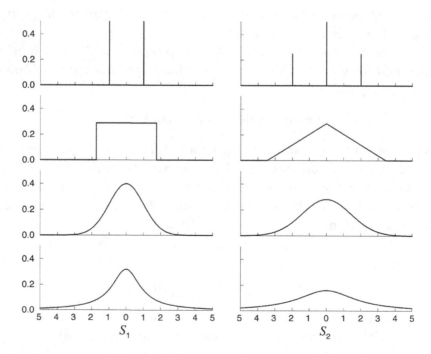

Fig. 3.2. Behavior of $P(S_n)$ for i.i.d. random variables with $n = 1, 2$ for the pdfs of Fig. 3.1.

Whereas all the distributions change as a function of n, a difference is observed between the first two and the Gaussian and Lorentzian distributions. The functions $P(S_n)$ for the delta and for the uniform distribution change both in scale and in functional form as n increases, while the Gaussian and the Lorentzian distributions do not change in shape but only in scale (they become broader when n increases). When the functional form of $P(S_n)$ is the same as the functional form of $P(x_i)$, the stochastic process is said to be *stable*. Thus Gaussian and Lorentzian processes are stable but, in general, stochastic processes are not.

3.3 Central limit theorem

Suppose that a random variable S_n is composed of many parts x_i, $S_n = \sum_{i=1}^{n} x_i$, such that each x_i is independent and with finite variance $E\{x_i\} = 0$, $E\{x_i^2\} = s_i^2$, and

$$\sigma_n^2 = E\{S_n^2\} = \sum_{i=1}^{n} s_i^2. \tag{3.9}$$

Suppose further that, when $\sigma_n \to \infty$, the Lindeberg condition [94] holds,

$$\frac{1}{\sigma_n^2} \sum_{i=1}^{n} E\{U_i^2\} \to 1, \tag{3.10}$$

where, for every $\epsilon > 0$, U_i is a truncated random variable that is equal to x_i when $|x_i| \leq \epsilon \sigma_n$ and zero otherwise. Then the central limit theorem (CLT) states that

$$\tilde{S}_n \equiv \frac{S_n}{\sigma_n} = \frac{x_1 + x_2 + \cdots + x_n}{\sigma \sqrt{n}} \tag{3.11}$$

is characterized by a Gaussian pdf with unit variance

$$P_G(\tilde{S}_n) = \frac{1}{\sqrt{2\pi}} \exp(-\tilde{S}_n^2/2). \tag{3.12}$$

A formal proof of the CLT is given in probability texts such as Feller [56].

Using two concrete examples, we 'illustrate' the main point of the theorem, the gradual convergence of $P(S_n)$ to the Gaussian shape when n increases. In our examples, we simulate the stochastic process S_n by assuming that x_i is characterized by (i) a double triangular $P(x_i)$ (Fig. 3.3) or (ii) a uniform $P(x_i)$ (Fig. 3.4). As expected, the $P(S_n)$ distribution broadens when n increases.

We emphasize the convergence to the Gaussian asymptotic distribution by plotting the pdf using scaled units, defining

$$\tilde{x} \equiv \frac{x}{n^{1/2}} \tag{3.13}$$

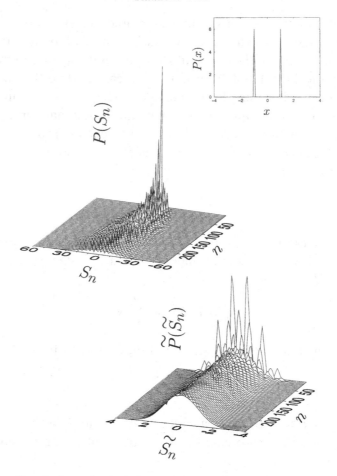

Fig. 3.3. Top: Simulation of $P(S_n)$ for n ranging from $n = 1$ to $n = 250$ for the case when $P(x)$ is a double triangular function (inset). Bottom: Same distribution using scaled units.

and

$$\tilde{P}(\tilde{x}) \equiv P(\tilde{x})n^{1/2}. \tag{3.14}$$

By analyzing the scaled pdfs $\tilde{P}(\tilde{x})$ observed at large values of n in Figs. 3.3 and 3.4, we note that the distributions rapidly converge to the functional form of the Gaussian of unit variance (shown as a smooth curve for large n).

We emphasize the fundamental hypothesis of the CLT. What is required is both independence and finite variance of the random variables x_i. When these conditions are not satisfied, other limit theorems must be considered (see Chapter 4).

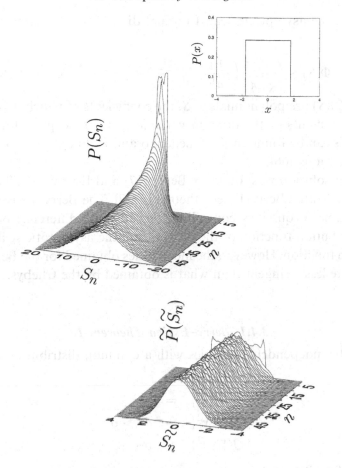

Fig. 3.4. Top: Simulation of $P(S_n)$ for n ranging from $n = 1$ to $n = 50$ for the case when $P(x)$ is uniformly distributed. Bottom: Same distribution in scaled units.

3.4 The speed of convergence

For independent random variables with finite variance, the CLT ensures that S_n will converge to a stochastic process with pdf

$$P_G(S_n) = \frac{1}{\sqrt{2\pi}\sigma_n} \exp(-S_n^2/2\sigma_n^2). \tag{3.15}$$

How fast is this convergence? Chebyshev considered this problem for a sum S_n of i.i.d. random variables x_i. He proved [30] that the scaled distribution function given by

$$F_n(S) \equiv \int_{-\infty}^{S} \tilde{P}(\tilde{S}_n) d\tilde{S}_n \tag{3.16}$$

differs from the asymptotic scaled normal distribution function $\Phi(S)$ by an amount

$$F_n(S) - \Phi(S) \sim \frac{e^{-S^2/2}}{\sqrt{2\pi}} \left(\frac{Q_1(S)}{n^{1/2}} + \frac{Q_2(S)}{n} + \cdots + \frac{Q_j(S)}{n^{j/2}} + \cdots \right), \qquad (3.17)$$

where the $Q_j(S)$ are polynomials in S, the coefficients of which depend on the first $j+2$ moments of the random variable $\{x_i\}$. The explicit form of these polynomials can be found in the Gnedenko and Kolmogorov monograph on limit distributions [66].

A simpler solution was found by Berry [17] and Esséen [51]. Their results are today called the Berry–Esséen theorems [57]. The Berry–Esséen theorems provide simple inequalities controlling the absolute difference between the scaled distribution function of the process and the asymptotic scaled normal distribution function. However, the inequalities obtained for the Berry–Esséen theorems are less stringent than what is obtained by the Chebyshev solution of Eq. (3.17).

3.4.1 Berry–Esséen Theorem 1

Let the x_i be independent variables with a common distribution function F such that

$$E\{x_i\} = 0 \qquad (3.18)$$
$$E\{x_i^2\} = \sigma^2 > 0 \qquad (3.19)$$
$$E\{|x_i|^3\} \equiv \rho < \infty. \qquad (3.20)$$

Then [57], for all S and n,

$$|F_n(S) - \Phi(S)| \le \frac{3\rho}{\sigma^3 \sqrt{n}}. \qquad (3.21)$$

The inequality (3.21) tells us that the convergence speed of the distribution function of $\tilde{S}_n$ to its asymptotic Gaussian shape is essentially controlled by the ratio of the third moment of the absolute value of x_i to the cube of the standard deviation of x_i.

3.4.2 Berry–Esséen Theorem 2

Theorem 2 is a generalization that considers random variables that might not be identically distributed. Let the x_i be independent variables such that

$$E\{x_i\} = 0 \qquad (3.22)$$
$$E\{x_i^2\} = \sigma_i^2 \qquad (3.23)$$
$$E\{|x_i|^3\} \equiv r_i, \qquad (3.24)$$

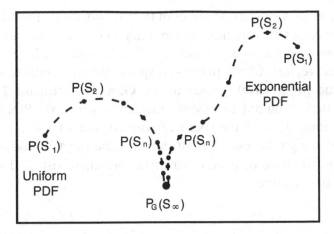

Fig. 3.5. Pictorial representation of the convergence to the Gaussian pdf $P_G(S_\infty)$ for the sum of i.i.d. finite variance random variables.

and define

$$s_n^2 \equiv \sigma_1^2 + \sigma_2^2 + \cdots + \sigma_n^2 \qquad (3.25)$$

and

$$\rho_n \equiv r_1 + r_2 + \cdots + r_n. \qquad (3.26)$$

Then [57] for all S and n,

$$|F_n(S) - \Phi(S)| \le 6\frac{\rho_n}{s_n^3}. \qquad (3.27)$$

3.5 Basin of attraction

The study of limit theorems uses the concept of the *basin of attraction* of a probability distribution. To introduce this concept, we focus our attention on the changes in the functional form of $P(S_n)$ that occur when n changes. We restrict our discussion to identically distributed random variables x_i. $P(S_1)$ then coincides with $P(x_i)$ and is characterized by the choices made in selecting the random variables x_i. When n increases, $P(S_n)$ changes its functional form and, if the hypotheses of the CLT are verified, assumes the Gaussian functional form for an asymptotically large value of n. The Gaussian pdf is an attractor (or fixed point) in the functional space of pdfs for all the pdfs that fulfill the requirements of the CLT. The set of such pdfs constitutes the basin of attraction of the Gaussian pdf.

In Fig. 3.5, we provide a pictorial representation of the motion of both the uniform and exponential $P(S_n)$ in the functional space of pdfs, and sketch the convergence to the Gaussian attractor of the two stochastic processes S_n. Both stochastic processes are obtained by summing n i.i.d. random variables x_i and y_i. The two processes x_i and y_i differ in their pdfs, indicated by their starting from different regions of the functional space. When n increases, both pdfs $P(S_n)$ become progressively closer to the Gaussian attractor $P_G(S_\infty)$. The number of steps required to observe the convergence of $P(S_n)$ to $P_G(S_\infty)$ provides an indication of the speed of convergence of the two families of processes. Although the Gaussian attractor is the most important attractor in the functional space of pdfs, other attractors also exist, and we consider them in the next chapter.

4
Lévy stochastic processes and limit theorems

In Chapter 3, we briefly introduced the concept of stable distribution, namely a specific type of distribution encountered in the sum of n i.i.d. random variables that has the property that it does not change its functional form for different values of n. In this chapter we consider the entire class of stable distributions and we discuss their principal properties.

4.1 Stable distributions

In §3.2 we stated that the Lorentzian and Gaussian distributions are stable. Here we provide a formal proof of this statement.

For Lorentzian random variables, the probability density function is

$$P(x) = \frac{\gamma}{\pi} \frac{1}{\gamma^2 + x^2}.$$ (4.1)

The Fourier transform of the pdf

$$\varphi(q) \equiv \int_{-\infty}^{+\infty} P(x)e^{iqx}dx$$ (4.2)

is called the characteristic function of the stochastic process. For the Lorentzian distribution, the integral is elementary. Substituting (4.1) into (4.2), we have

$$\varphi(q) = e^{-\gamma|q|}.$$ (4.3)

The convolution theorem states that the Fourier transform of a convolution of two functions is the product of the Fourier transforms of the two functions,

$$\mathscr{F}\left[f(x) \bigotimes g(x)\right] = \mathscr{F}[f(x)]\mathscr{F}[g(x)] = F(q)G(q).$$ (4.4)

For i.i.d. random variables,

$$S_2 = x_1 + x_2.$$ (4.5)

23

The pdf $P_2(S_2)$ of the sum of two i.i.d. random variables is given by the convolution of the two pdfs of each random variable

$$P_2(S_2) = P(x_1) \otimes P(x_2), \tag{4.6}$$

The convolution theorem then implies that the characteristic function $\varphi_2(q)$ of S_2 is given by

$$\varphi_2(q) = [\varphi(q)]^2. \tag{4.7}$$

In the general case,

$$P_n(S_n) = P(x_1) \otimes P(x_2) \otimes \cdots \otimes P(x_n), \tag{4.8}$$

where S_n is defined by (3.1). Hence

$$\varphi_n(q) = [\varphi(q)]^n. \tag{4.9}$$

The utility of the characteristic function approach can be illustrated by obtaining the pdf for the sum S_2 of two i.i.d. random variables, each of which obeys (4.1). Applying (4.6) would be cumbersome, while the characteristic function approach is quite direct, since for the Lorentzian distribution,

$$\varphi_2(q) = e^{-2|q|\gamma}. \tag{4.10}$$

By performing the inverse Fourier transform

$$P(x) = \frac{1}{2\pi} \int_{-\infty}^{+\infty} \varphi(q) e^{-iqx} dq, \tag{4.11}$$

we obtain the probability density function

$$P_2(S_2) = \frac{2\gamma}{\pi} \frac{1}{4\gamma^2 + x^2}. \tag{4.12}$$

The functional form of $P_2(S_2)$, and more generally of $P_n(S_n)$, is Lorentzian. Hence a Lorentzian distribution is a stable distribution.

For Gaussian random variables, the analog of (4.1) is the pdf

$$P(x) = \frac{1}{\sqrt{2\pi}\sigma} e^{-x^2/2\sigma^2}. \tag{4.13}$$

The characteristic function is

$$\varphi(q) = e^{-(\sigma^2/2)q^2} = e^{-\gamma q^2}, \tag{4.14}$$

where $\gamma \equiv \sigma^2/2$. Hence from (4.7),

$$\varphi_2(q) = e^{-2\gamma q^2}. \tag{4.15}$$

By performing the inverse Fourier transform, we obtain

$$P_2(S_2) = \frac{1}{\sqrt{8\pi\gamma}} e^{-x^2/8\gamma}. \tag{4.16}$$

Thus the Gaussian distribution is also a stable distribution. Writing (4.16) in the form

$$P_2(S_2) = \frac{1}{\sqrt{2\pi}(\sqrt{2}\sigma)} e^{-x^2/2(\sqrt{2}\sigma)^2}, \tag{4.17}$$

we find

$$\sigma_2 = \sqrt{2}\sigma. \tag{4.18}$$

We have verified that two stable stochastic processes exist: Lorentzian and Gaussian. The characteristic functions of both processes have the same functional form

$$\varphi(q) = e^{-\gamma|q|^\alpha}, \tag{4.19}$$

where $\alpha = 1$ for the Lorentzian from (4.3), and $\alpha = 2$ for the Gaussian from (4.15).

Lévy [92] and Khintchine [80] solved the general problem of determining the entire class of stable distributions. They found that the most general form of a characteristic function of a stable process is

$$\ln \varphi(q) = \begin{cases} i\mu q - \gamma|q|^\alpha \left[1 - i\beta \frac{q}{|q|} \tan\left(\frac{\pi}{2}\alpha\right) \right] & [\alpha \neq 1] \\ i\mu q - \gamma|q| \left[1 + i\beta \frac{q}{|q|} \frac{2}{\pi} \ln|q| \right] & [\alpha = 1] \end{cases}, \tag{4.20}$$

where $0 < \alpha \leq 2$, γ is a positive scale factor, μ is any real number, and β is an asymmetry parameter ranging from -1 to 1.

The analytical form of the Lévy stable distribution is known only for a few values of α and β:

- $\alpha = 1/2$, $\beta = 1$ (Lévy–Smirnov)
- $\alpha = 1$, $\beta = 0$ (Lorentzian)
- $\alpha = 2$ (Gaussian)

Henceforth we consider here only the symmetric stable distribution ($\beta = 0$) with a zero mean ($\mu = 0$). Under these assumptions, the characteristic function assumes the form of Eq. (4.19). The symmetric stable distribution of index α and scale factor γ is, from (4.20) and (4.11),

$$P_L(x) \equiv \frac{1}{\pi} \int_0^\infty e^{-\gamma|q|^\alpha} \cos(qx)dq. \tag{4.21}$$

For $\gamma = 1$, a series expansion valid for large arguments ($|x| \gg 0$) is [16]

$$P_L(|x|) = -\frac{1}{\pi} \sum_{k=1}^{n} \frac{(-1)^k}{k!} \frac{\Gamma(\alpha k + 1)}{|x|^{\alpha k + 1}} \sin\left[\frac{k\pi\alpha}{2}\right] + R(|x|), \qquad (4.22)$$

where $\Gamma(x)$ is the Euler Γ function and

$$R(|x|) = \mathcal{O}(|x|^{-\alpha(n+1)-1}). \qquad (4.23)$$

From (4.22) we find the asymptotic approximation of a stable distribution of index α valid for large values of $|x|$,

$$P_L(|x|) \sim \frac{\Gamma(1+\alpha)\sin(\pi\alpha/2)}{\pi|x|^{1+\alpha}} \sim |x|^{-(1+\alpha)}. \qquad (4.24)$$

The asymptotic behavior for large values of x is a power-law behavior, a property with deep consequences for the moments of the distribution. Specifically, $E\{|x|^n\}$ diverges for $n \geq \alpha$ when $\alpha < 2$. In particular, all Lévy stable processes with $\alpha < 2$ have *infinite* variance. Thus non-Gaussian stable stochastic processes do not have a characteristic scale – the variance is infinite!

4.2 Scaling and self-similarity

We have seen that Lévy distributions are stable. In this section, we will argue that these stable distributions are also self-similar. How do we rescale a non-Gaussian stable distribution to reveal its self-similarity? One way is to consider the 'probability of return to the origin' $P(S_n = 0)$, which we obtain by starting from the characteristic function

$$\varphi_n(q) = e^{-n\gamma|q|^\alpha}. \qquad (4.25)$$

From (4.11),

$$P(S_n) = \frac{1}{\pi} \int_0^\infty e^{-n\gamma|q|^\alpha} \cos(qS_n) dq. \qquad (4.26)$$

Hence

$$P(S_n = 0) = \frac{1}{\pi} \int_0^\infty e^{-n\gamma|q|^\alpha} dq = \frac{\Gamma(1/\alpha)}{\pi\alpha(\gamma n)^{1/\alpha}}. \qquad (4.27)$$

The $P(S_n)$ distribution is properly rescaled by defining

$$\tilde{P}(\tilde{S}_n) \equiv P(S_n)n^{1/\alpha}. \qquad (4.28)$$

The normalization

$$\int_{-\infty}^{+\infty} \tilde{P}(\tilde{S}_n) d\tilde{S}_n = 1, \qquad (4.29)$$

is assured if

$$\tilde{S}_n \equiv \frac{S_n}{n^{1/\alpha}}. \tag{4.30}$$

When $\alpha = 2$, the scaling relations coincide with what we used for a Gaussian process in Chapter 3, namely Eqs. (3.13) and (3.14).

4.3 Limit theorem for stable distributions

In the previous chapter, we discussed the central limit theorem and we noted that the Gaussian distribution is an attractor in the functional space of pdfs. The Gaussian distribution is a peculiar stable distribution; it is the only stable distribution having all its moments finite. It is then natural to ask if non-Gaussian stable distributions are also attractors in the functional space of pdfs. The answer is affirmative. There exists a limit theorem [65, 66] stating that the pdf of a sum of n i.i.d. random variables x_i converges, in probability, to a stable distribution under certain conditions on the pdf of the random variable x_i. Consider the stochastic process $S_n = \sum_{i=1}^{n} x_i$, with x_i being i.i.d. random variables. Suppose

$$P(x_i) \sim \begin{cases} C_-|x_i|^{-(1+\alpha)} & \text{as } x \to -\infty \\ C_+|x_i|^{-(1+\alpha)} & \text{as } x \to +\infty \end{cases}, \tag{4.31}$$

and

$$\beta \equiv \frac{C_+ - C_-}{C_+ + C_-}. \tag{4.32}$$

Then $\tilde{P}(\tilde{S}_n)$ approaches a stable non-Gaussian distribution $P_L(x)$ of index α and asymmetry parameter β, and $P(S_n)$ belongs to the attraction basin of $P_L(x)$.

Since α is a continuous parameter over the range $0 < \alpha \le 2$, an infinite number of attractors is present in the functional space of pdfs. They comprise the set of all the stable distributions. Figure 4.1 shows schematically several such attractors, and also the convergence of a certain number of stochastic processes to the asymptotic attracting pdf. An important difference is observed between the Gaussian attractor and stable non-Gaussian attractors: finite variance random variables are present in the Gaussian basin of attraction, whereas random variables with infinite variance are present in the basins of attraction of stable non-Gaussian distributions. We have seen that stochastic processes with infinite variance are characterized by distributions with power-law tails. Hence such distributions with power-law tails are present in the stable non-Gaussian basins of attraction.

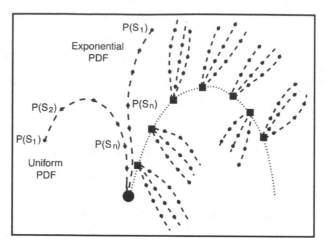

Fig. 4.1. Pictorial representation of the convergence process (in probability) to some of the stable attractors of the sum of i.i.d. random variables. The black circle is the Gaussian attractor and the black squares the Lévy stable non-Gaussian attractors characterized by different values of the index α.

4.4 Power-law distributions

Are power-law distributions meaningful or meaningless? Mathematically they are meaningful, despite the presence of diverging moments. Physically, they are meaningless for finite ('isolated') systems. For example, an infinite second moment in the formalism of equilibrium statistical mechanics would imply an infinite temperature.

What about open ('non-isolated') systems? Indeed, Bernoulli considered random variables with infinite expectations in describing a fair game, the St Petersburg paradox, while Pareto found power-law distributions empirically in the distribution of incomes. Mandelbrot used power-law distributions in describing economic and physical systems.

Power-law distributions are counterintuitive because they lack a characteristic scale. More generally, examples of random variables with infinite expectations were treated as paradoxes before the work of Lévy. A celebrated example is the St Petersburg paradox. N. Bernoulli introduced the game in the early 1700s and D. Bernoulli wrote about it in the *Commentary of the St Petersburg Academy* [56].

4.4.1 The St Petersburg paradox

A banker flips a coin $n + 1$ times. The player wins 2^{n-1} coins if n tails occur before the first head. The outcomes are made clear in the following chart:

n	coins won	probability	expected winnings
1	1	1/2	$1 \times 1/2$
2	2	1/4	$2 \times 1/4$
3	4	1/8	$4 \times 1/8$
$\vdots$	$\vdots$	$\vdots$	$\vdots$
n	2^{n-1}	$1/2^n$	$1/2$
$\vdots$	$\vdots$	$\vdots$	$\vdots$

The cumulative expected win is $1/2 + 1/2 + \cdots = \infty$. How many coins must the player risk in order to play? To determine the fair 'ante', each party must decide how much he is willing to gamble. Specifically, the banker asks for his expected loss – it is an infinite number of coins. The player disagrees because he assumes he will not win an infinite number of coins with probability one (two coins or fewer with probability 3/4, four coins or fewer with probability 7/8, and so on). The two parties cannot come to an agreement. Why? The 'modern' answer is that they are trying to determine a characteristic scale for a problem that has no characteristic scale.

4.4.2 Power laws in finite systems

Today, power-law distributions are used in the description of open systems. However, the scaling observed is often limited by finite size effects or some other limitation intrinsic to the system. A good example of the fruitful use of power laws and of the difficulties related to their use is provided by critical phenomena [147]. Power-law correlation functions are observed in the critical state of an infinite system, but if the system is finite, the finiteness limits the range within which a power-law behavior is observed. In spite of this limitation, the introduction and the use of the concept of scaling – which is related to the power-law nature of correlation – is crucial for the understanding of critical phenomena even when finite systems are considered [59].

4.5 Price change statistics

In this book, we are considering the limit theorems of probability theory to have a theoretical framework that tells us what kind of distribution we should expect for price changes in financial markets. Stable non-Gaussian distributions are of interest because they obey limit theorems. However, we

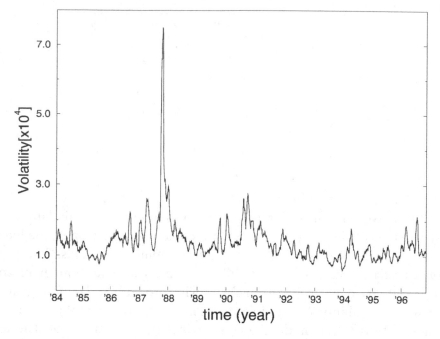

Fig. 4.2. Monthly volatility of the S&P 500 index measured for the 13-year period January 1984 to December 1996. Courtesy of P. Gopikrishnan.

should not expect to observe price change distributions that are stable. The reason is related to the hypotheses underlying the limit theorem for stable distributions: The random variables $\{x_i\}$ are (i) pairwise-independent and (ii) identically distributed. Hypothesis (i) has been well verified for time horizons ranging from a few minutes to several years. However, hypothesis (ii) is not generally verified by empirical observation because, e.g., the standard deviation of price changes is strongly time-dependent. This phenomenon is known in finance as time-dependent volatility [143] (an example is shown in Fig. 4.2).

A more appropriate limit theorem is one based only on the assumption that random variables x_i are independent but not necessarily identically distributed. A limit theorem valid for a sum S_n of independent random variables $\{x_i\}$ was first presented by Bawly and Khintchine [66, 81], who considered the class of limit laws for the sum S_n of n independent infinitesimal random variables. Infinitesimal is used here as a technical term meaning that in the sum S_n there is no single stochastic variable x_i that dominates the sum. Then the Khintchine theorem states that it is necessary and sufficient that $F_n(S)$, the limit distribution function, be infinitely divisible.

4.6 Infinitely divisible random processes

A random process y is infinitely divisible if, for every natural number k, it can be represented as the sum of k i.i.d. random variables $\{x_i\}$. The distribution function $F(y)$ is infinitely divisible if and only if the characteristic function $\varphi(q)$ is, for every natural number k, the kth power of some characteristic function $\varphi_k(q)$. In formal terms

$$\varphi(q) = [\varphi_k(q)]^k, \tag{4.33}$$

with the requirements (i) $\varphi_k(0) = 1$ and (ii) $\varphi_k(q)$ is continuous.

4.6.1 Stable processes

A normally distributed random variable $\{y\}$ is infinitely divisible because, from (4.14),

$$\varphi(q) = \exp\left[i\mu q - \frac{\sigma^2}{2}q^2\right], \tag{4.34}$$

so a solution of the functional equation (4.33) is

$$\varphi_k(q) = \exp\left[\frac{i\mu q}{k} - \frac{\sigma^2}{2k}q^2\right]. \tag{4.35}$$

A symmetric stable random variable is infinitely divisible. In fact, from (4.19)

$$\varphi(q) = \exp[i\mu q - \gamma|q|^\alpha], \tag{4.36}$$

so

$$\varphi_k(q) = \exp\left[\frac{i\mu q}{k} - \frac{\gamma}{k}|q|^\alpha\right]. \tag{4.37}$$

4.6.2 Poisson process

The Poisson process $P(m; \lambda) = e^{-\lambda}(\lambda^m/m!)$, with $m = 0, 1, \ldots, n$, has a characteristic function

$$\varphi(q) = \exp[\lambda(e^{iq} - 1)], \tag{4.38}$$

so, from (4.33),

$$\varphi_k(q) = \exp\left[\frac{\lambda}{k}(e^{iq} - 1)\right]. \tag{4.39}$$

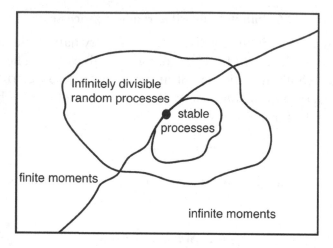

Fig. 4.3. Illustrative scheme of the classes of random processes discussed in this chapter. The solid circle denotes the stable Gaussian process.

4.6.3 Gamma distributed random variables

The Gamma distribution has pdf

$$P(x) = \frac{e^{-x}\, x^{v-1}}{\Gamma(v)}. \tag{4.40}$$

For $x \geq 0$ and $0 < v < \infty$, the characteristic function is

$$\varphi(q) = (1 - iq)^{-v}, \tag{4.41}$$

so, from (4.33),

$$\varphi_k(q) = (1 - iq)^{-v/k}. \tag{4.42}$$

4.6.4 Uniformly distributed random variables

The class of infinitely divisible stochastic processes is large, but there are several stochastic processes that are not infinitely divisible. One example is a random process with a uniform pdf

$$P(x) = \begin{cases} 0 & x < -\ell \\ 1/2\ell & -\ell \leq x \leq \ell \\ 0 & x > \ell \end{cases}. \tag{4.43}$$

In this case, the characteristic function is

$$\varphi(q) = \frac{\sin(q\ell)}{q\ell},$$ (4.44)

and the process is not infinitely divisible because the kth root does not exist.

4.7 Summary

In Fig. 4.3 we provide a schematic illustration of some important classes of stochastic processes discussed in this chapter.

The class of infinitely divisible random processes is a large class that includes the class of stable random processes. Infinitely divisible random processes may have finite or infinite variance. Stable non-Gaussian random processes have infinite variance, whereas the Gaussian process is the only stable process with finite variance.

Empirical observations, together with limit theorems of probability theory, allow one to conclude that the pdf of price changes must progressively converge to an infinitely divisible pdf for long time horizons. Hence the Khintchine limit theorem ensures that for large values of n, the price change distribution is well defined, in spite of the fact that the price change stochastic process $Z(t)$ at a time t may be characterized by parameters and functional forms that are t-dependent. Moreover, the Khintchine theorem states that the distribution $P(Z)$ is close to an infinitely divisible pdf and the degree of convergence increases when n increases. Hence a long time horizon pdf of price changes can be considered in terms of a sum of i.i.d. random variables. Even in the presence of volatility fluctuations, it is possible to model price changes in terms of newly defined i.i.d. random variables. These are the variables defined by Eq. (4.33). One must keep in mind that the information extracted from this i.i.d. random process applies to pdfs for long time horizons, and not to local time scales.

5

Scales in financial data

A truly gargantuan quantity of financial data is currently being recorded and stored in computers. Indeed, nowadays every transaction of every financial market in the entire world is recorded somewhere. The nature and format of these data depend upon the financial asset in question and on the particular institution collecting the data. Data have been recorded

- on a daily basis since the 19th century (Fig. 5.1, for an example),
- with a sampling rate of 1 min or less since 1984 (Fig. 5.2), and
- transaction-by-transaction ('tick-by-tick') since 1993 (Fig. 5.3).

Statistical analyses of financial data have been performed since the recording activity started. Since the 1950s, when computer data processing became

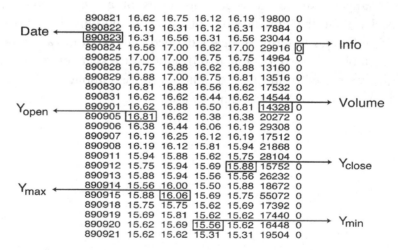

Fig. 5.1. Daily data on the prices of Coca Cola Co. stock. Records show the date, the open price, the maximum and the minimum price during the day, the closing price, the volume traded during the day, and additional information on the record.

34

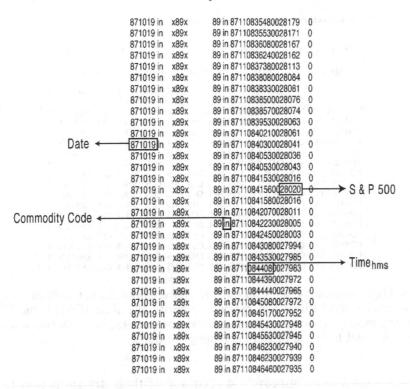

Fig. 5.2. High-frequency records of the S&P 500 index. The data contain information on the value of the index at each time for which it is calculated.

available, statistical analysis has progressively involved a larger and larger number of financial records. For example, Mandelbrot's 1963 cotton-price study [102] analyzed $\approx 2 \times 10^3$ records, the 1995 study [111] of the Standard & Poor's 500 index analyzed $\approx 5 \times 10^5$ records, and a recent tick-by-tick study [67] used 4×10^7 relative price changes for the 1,000 largest companies traded in the New York Stock Exchange.

Statistical analyses of market data are essential, both for the fundamental reason of understanding market dynamics and for applied reasons related to the key problems of option pricing and portfolio management. In this chapter we consider some peculiarities of financial data, scales and units. Indeed, the role of scales and reference units in finance and physics is rather different, and we discuss this difference in detail.

5.1 Price scales in financial markets

In physics, the problem of reference units is considered basic to all experimental and theoretical work. Efforts are continually made to find the optimal

Tick extraction for JPY DEM
_____

Filtering: O&A standard filter for historical tests

date	time	price	country	filter
			city	Good (1) or Bad (0)

CCYY-MM-DD (GMT)	bid	ask		bank	
1992-10-01 00:01:02	84.90	84.95	702 01	0041	1
1992-10-01 00:05:18	84.93	84.96	036 02	0032	1
1992-10-01 00:07:10	84.92	84.97	344 01	0136	1
1992-10-01 00:11:32	84.95	85.00	344 01	0089	1
1992-10-01 00:16:34	84.96	85.01	344 01	0136	1
1992-10-01 00:16:50	84.98	85.03	344 01	0136	1
1992-10-01 00:17:26	85.00	85.05	344 01	0136	1
1992-10-01 00:18:16	84.95	85.00	392 01	0033	1
1992-10-01 00:20:00	84.96	85.01	344 01	0136	1
1992-10-01 00:20:18	84.95	85.00	344 01	0437	1
1992-10-01 00:21:06	84.95	85.00	702 01	0055	1
1992-10-01 00:23:00	85.00	85.05	344 01	0136	1
1992-10-01 00:23:18	85.02	85.07	344 01	0136	1
1992-10-01 00:23:36	85.00	85.05	702 01	0055	1
1992-10-01 00:23:56	84.96	85.01	344 01	0136	1

Fig. 5.3. High-frequency quotes on the foreign exchange market, collected by Olsen & Associates Corporation. The records comprise the time (GMT), the bid, the ask on Japanese yen/German Deutschmark transactions, and information on the country, city, and financial institution issuing the quote.

reference units and to improve the accuracy of their determination [33, 40]. A branch of physics, metrology, is exclusively devoted to this task, and large specialized institutions in metrology exist all over the world. In finance, almost the opposite is the case. The scales used are often given in units (currencies) that are themselves fluctuating in time and transactions occur at random times with random intensities. For this reason, great care must be taken in the selection of the most appropriate variable to be studied, taking into account the implicit assumptions associated with each possible choice.

Here we first consider the problem of price scales. In the next section we consider the problem of time scales.

The price unit of financial goods is usually the currency of the country in which the particular financial market is located. The value of the currency is not constant in time. A currency can change its value because of

- inflation,
- economic growth or economic recession, and
- fluctuations in the global currency market.

Examples of some macroeconomic records are given in Figs. 5.4 and 5.5. In Fig. 5.4, we show a table of the annual percent changes of the gross domestic product of several industrial countries at constant 1980 currency

GDP at Constant Prices

1978	1979	1980	1981	1982	1983	1984	1985	1986	1987		99bp x
4.1	3.7	2.2	1.7	1.2	2.1	4.1	2.9	....	....	World	001
4.0	3.2	1.4	1.5	1.1	2.6	4.5	3.1	3.0	3.0	Industrial Countries	110
5.2	2.1	-.2	2.0	-2.5	3.7	6.6	3.0	3.8	3.1	United States	111
4.2	3.7	1.5	3.0	-3.4	3.7	6.1	4.3	3.0	....	Canada*	156
.8	3.4	3.4	2.0	3.1	.4	6.7	5.5	1.8	4.4	Australia	193
5.2	5.3	4.3	3.7	3.1	3.2	5.1	4.7	2.5	4.4	Japan*	158
—	2.7	.7	1.4	3.1	.1	6.6	1.5	....	....	New Zealand	196
.5	4.7	3.0	-.1	1.1	2.2	1.4	2.8	1.7	1.3	Austria	122
2.9	2.2	4.1	-1.3	1.5	.1	2.0	1.4	2.4	1.7	Belgium	124
1.5	3.5	-.4	-.9	3.0	2.5	4.4	4.2	3.3	-1.0	Denmark	128
2.2	7.3	5.4	1.6	3.6	3.0	3.3	3.5	2.4	....	Finland	172
3.3	3.2	1.6	1.2	2.5	.7	1.3	1.7	2.1	2.2	France	132
3.0	4.1	1.7	.2	11.7	1.5	2.8	2.1	2.6	1.8	Germany	134
5.9	5.0	-4.2	1.6	-1.5	-5.5	2.7	10.1	6.3	5.5	Iceland*	176
7.2	3.1	3.1	3.3	2.3	-1.1	3.8	1.1	-.3	....	Ireland	178
2.7	4.9	3.9	1.1	.2	1.0	3.2	2.8	2.9	3.1	Italy	136
4.7	4.0	2.9	.5	1.5	2.4	5.7	3.9	3.4	....	Luxembourg	137
2.5	2.4	.9	-.7	-1.4	1.4	3.2	2.3	2.4	2.2	Netherlands	138
4.5	5.1	14.2	.9	.3	4.6	15.7	5.3	14.2	1.3	Norway	142
1.8	.2	1.5	-.2	1.2	1.8	1.9	2.1	3.6	....	Spain	184
1.8	3.8	1.7	-.3	.8	2.4	3.9	2.1	1.2	2.8	Sweden	144
.4	2.5	4.6	1.5	-1.1	.7	2.1	3.7	2.8	....	Switzerland	146
3.9	2.1	-2.1	-.9	1.1	3.5	2.1	3.9	2.9	3.6	United Kingdom	112

Fig. 5.4. Annual percent change of the gross domestic product of several countries over a 10-year period; data are obtained from International Financial Statistics (International Monetary Fund, 1988), page 165.

	Jan	Feb	Mar	Apr	May	June	July	Aug	Sept	Oct	Nov	Dec
1972	49.9	50.2	50.2	50.4	50.5	50.6	50.9	50.9	51.1	51.3	51.4	51.6
1973	51.7	52.1	52.6	53.0	53.3	53.6	53.8	54.0	54.9	55.3	55.8	56.1
1974	56.6	57.3	58.0	58.3	59.0	59.5	60.0	60.7	61.5	62.0	62.5	63.0
1975	63.2	63.7	63.9	64.3	64.5	65.1	65.8	66.0	66.3	66.7	67.1	67.4
1976	67.5	67.7	67.9	68.2	68.6	68.9	69.3	69.7	69.9	70.2	70.4	70.6
1977	71.0	71.8	72.2	72.8	73.2	73.7	74.0	74.3	74.6	74.8	75.1	75.4
1978	75.9	76.3	76.9	77.6	78.3	79.1	79.7	80.1	80.8	81.4	81.8	82.2
1979	82.9	83.9	84.7	85.7	86.8	87.8	88.7	89.6	90.5	91.3	92.2	93.2
1980	94.5	95.8	97.2	98.3	99.2	100.3	100.4	101.1	102.0	102.9	103.8	104.7
1981	105.6	106.6	107.4	108.1	109.0	109.9	111.2	112.0	113.2	113.4	113.7	114.1
1982	114.5	114.8	114.7	115.2	116.3	117.7	118.4	118.6	118.8	119.2	119.0	118.5
1983	118.8	118.8	118.9	119.7	120.4	120.8	121.3	121.7	122.3	122.6	122.8	123.0
1984	123.7	124.2	124.5	125.1	125.5	125.9	126.3	126.8	127.4	127.8	127.8	127.8
1985	128.1	128.6	129.2	129.7	130.2	130.6	130.8	131.1	131.5	131.9	132.3	132.7
1986	133.1	132.7	132.1	131.8	132.2	132.9	132.9	133.1	133.8	133.9	134.0	134.2

Fig. 5.5. Monthly consumer price index in the United States during the 15-year period 1972 to 1986, normalized to the value of 100 USD for the year 1980. Data from International Financial Statistics, Supplement on Price Statistics (International Monetary Fund, 1986), page 70.

values. The economic growth of a country is itself a random variable. In Fig. 5.5, the monthly values of the US consumer price index are shown for the period 1972 to 1986. Also for this economic indicator, periods of high levels of inflation alternate with periods of low levels. For economic indicators, many stochastic descriptions have been developed.

Let us define $Y(t)$ as the price of a financial asset at time t. Which is the appropriate stochastic variable for us to investigate? Different choices are possible and each has its merits and its problems. Below, we discuss the most common choices.

(i) One can investigate price changes,

$$Z(t) \equiv Y(t + \Delta t) - Y(t). \tag{5.1}$$

The merit of this approach is that nonlinear or stochastic transformations are not needed. The problem is that this definition is seriously affected by changes in scale.

(ii) Alternatively, one can analyze deflated, or discounted, price changes,

$$Z_D(t) \equiv [Y(t + \Delta t) - Y(t)]D(t), \tag{5.2}$$

where $D(t)$ can be a deflation factor, or a discounting factor. The merits of this approach are that (i) nonlinear transformations are not needed and (ii) prices are given in terms of 'constant' money – the gains possible with riskless investments are accounted for by the factor $D(t)$. The problem is that deflators and discounting factors are unpredictable over the long term, and there is no unique choice for $D(t)$.

(iii) One can analyze returns, defined as

$$R(t) \equiv \frac{Y(t + \Delta t) - Y(t)}{Y(t)} = \frac{Z(t)}{Y(t)}. \tag{5.3}$$

The merit of this approach is that returns provide a direct percentage of gain or loss in a given time period. The problem is that returns are sensitive to scale changes for long time horizons.

(iv) One can study the successive differences of the natural logarithm of price,

$$S(t) \equiv \ln Y(t + \Delta t) - \ln Y(t). \tag{5.4}$$

The merit of this approach is that the average correction of scale changes is incorporated without requiring deflators or discounting factors. The problems are (a) the correction of scale change would be correct only if the growth rate of the economy were constant, but the growth rate generally fluctuates, and these fluctuations are not incorporated into definition (5.4), and (b) a nonlinear transformation is used, and nonlinearity strongly affects the statistical properties of a stochastic process. Note that the information carried by $S(t)$ mixes features of the dynamics of the financial asset together with aspects involving fluctuations of macroeconomic indicators.

The analysis of high-frequency financial data has become widespread in research institutes and financial institutions, and it is worthwhile to consider how the above definitions are interrelated in the high-frequency regime. From (5.4) and (5.1),

$$S(t) = \ln \frac{Y(t + \Delta t)}{Y(t)} = \ln \left[1 + \frac{Z(t)}{Y(t)}\right]. \tag{5.5}$$

For high-frequency data, Δt is small and $|Z(t)| \ll Y(t)$. Hence

$$S(t) = \ln\left[1 + \frac{Z(t)}{Y(t)}\right] \approx \frac{Z(t)}{Y(t)} = R(t). \tag{5.6}$$

Since $Z(t)$ is a fast variable whereas $Y(t)$ is a slow variable,

$$R(t) \approx C_1 Z(t), \tag{5.7}$$

where the time dependence of C_1 is negligible. Moreover, if the total investigated time period is not too long, $D(t) \approx 1$, so

$$Z(t) \approx Z_D(t). \tag{5.8}$$

To summarize, for high-frequency data and for investigations limited to a short time period in a time of low inflation, all four commonly used indicators are approximately equal:

$$S(t) \approx R(t) \approx C_1 Z(t) \approx Z_D(t). \tag{5.9}$$

However, for investigations over longer time periods, a choice must be made. The most commonly studied functions are $S(t)$ and $R(t)$.

5.2 Time scales in financial markets

Next we consider the problem of choosing the appropriate time scale to use for analyzing market data. Possible candidates for the 'correct' time scale include:

- the physical time,
- the trading (or market) time, or
- the number of transactions.

An indisputable choice is not available. As in the case of price scale unit, all the definitions have merits and all have problems. When examining price changes that take place when transactions occur, it is worth noting that each transaction occurring at a random time (see Fig. 5.6) involves a random variable, the volume, of the traded financial good.

Physical time is well defined, but stock exchanges close at night, over weekends, and during holidays. A similar limitation is also present in a global market such as the foreign exchange market. Although this market is active 24 hours per day, the social organization of business and the presence of biological cycles force the market activity to have temporal constraints in each financial region of the world. With the choice of a *physical time*, we do

Scales in financial data

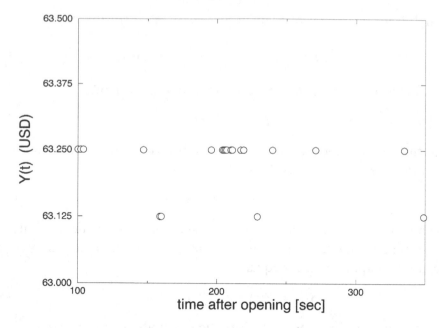

Fig. 5.6. Price change during the day of 3 January 1994 of an Exxon stock traded in the New York Stock Exchange. The price is recorded when a transaction occurs, and transactions occur randomly in time.

not know how to model the stochastic dynamics of prices and the arrival of information during hours in which the market is closed.

Trading time is well defined in stock exchanges – it is the time that elapses during open market hours. In the foreign exchange market, it coincides with the physical time. Empirical studies have tried to determine the variance of log price changes observed from closure to closure in financial markets. These studies show that the variance determined by considering closure values of successive days is only approximately 20% lower than the variance determined by considering closure values across weekends [52, 60]. This empirical evidence supports the choice of using trading time in the modeling of price dynamics. Indeed, the trading time is the most common choice in research studies and in the studies performed for the determination of volatility in option pricing. However, problems also arise with this definition. Specifically,

(i) information affecting the dynamics of the price of a financial asset can be released while the market is closed (or its activity is negligible in a given financial area),

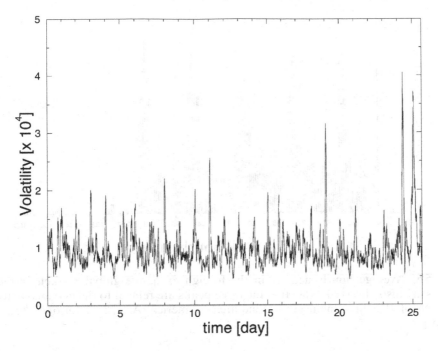

Fig. 5.7. Volatility (to be discussed in Chapter 7) of the S&P 500 high-frequency data. A daily cycle with a period of 6.5 trading hours is clearly observed in the time evolution.

(ii) in high-frequency analyses overnight price changes are treated as short-time price changes, and

(iii) the market activity is implicitly assumed to be uniform during market hours.

This last assumption is not verified by empirical analyses. Trading activity is not uniform during trading hours, either in terms of volume or in number of contracts. Rather, a daily cycle is observed in market data: the volatility is higher at the opening and closing hours, and usually the lowest value of the day occurs during the middle hours. As an example, we show in Fig. 5.7 the intraday 1-minute volatility of the S&P 500 index determined each trading hour. Clearly seen is a daily cycle with a period of 6.5 trading hours.

An analogous, almost periodic behavior is observed in the average activity of the foreign exchange market (Fig. 5.8). In this case, the three different peaks of the intraday cycle observed in the volatility of price changes reflect the daily peak activity in three different regions of the world – Asia, Europe, and the Americas [41].

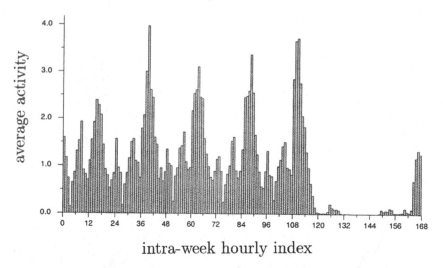

intra-week hourly index

Fig. 5.8. Average hourly activity in the foreign exchange global market. Intraday cycles are also observed. Note that the three peaks are related to the maximal activity in each of the three main geographic areas, America, Asia, and Europe. Adapted from [41].

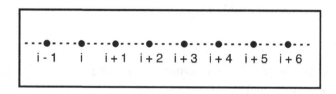

Fig. 5.9. Schematic illustration of the occurrence of successive transactions in transaction units.

One can explore other definitions of temporal activity that are not affected by the fact that trading activity is not uniform in time. One definition concerns the time index of the number of effective transactions occurring in the market for a given financial asset. The use of this definition is not easy because tick-by-tick data are necessary to perform a statistical analysis in terms of such a time index. However, such an analysis is possible today because tick-by-tick data are available, at least for some financial markets.

If 'time' is defined in terms of the number of transactions (Fig. 5.9), then one source of randomness observed in financial markets is eliminated, specifically the time elapsing between transactions. However, the second source of randomness, the volume of the transaction, still remains.

5.3 Summary

It is not straightforward to select the price function and the time reference frame to be used in the analysis and modeling of the stochastic dynamics of a price. Several choices are possible, each based on explicit or implicit assumptions that may or may not be verifiable for an asset in a given period of time. Empirical analyses are often performed with slightly different (i) definitions of the variables investigated, e.g., returns and log price differences, (ii) periods of time analyzed, and (iii) frequency of recorded data. Results are sensitive to these choices, so particular care must be taken when we compare results obtained by different researchers for different financial goods under different time conditions. Perhaps this is one of the reasons why the complete characterization of the statistical properties of price changes is still lacking, despite a large number of empirical analyses.

6
Stationarity and time correlation

In this chapter we consider the degree of stationarity observed in time series of price changes in financial markets. We discuss various definitions of stationarity, and consider which of them best applies to financial data. We take a similar approach concerning time correlation, namely we first discuss the classes of correlation of short-range and long-range correlated stochastic processes, and then we present and discuss some empirical studies of financial data. When the stochastic variables are independent, stationarity implies that the stochastic process $x(t)$ is independent identically distributed.

The statistical observables characterizing a stochastic process can be written in terms of nth-order statistical properties. The case $n = 1$ is sufficient to define the mean,

$$E\{x(t)\} \equiv \int_{-\infty}^{\infty} x f(x, t) dx, \qquad (6.1)$$

where $f(x, t)$ gives the probability density of observing the random value x at time t. The case $n = 2$ is used to define the autocorrelation function

$$E\{x(t_1)x(t_2)\} \equiv \int_{-\infty}^{\infty} \int_{-\infty}^{\infty} x_1 x_2 f(x_1, x_2; t_1, t_2) dx_1 dx_2, \qquad (6.2)$$

where $f(x_1, x_2; t_1, t_2)$ is the joint probability density that x_1 is observed at time t_1 and x_2 is observed at time t_2. To fully characterize the statistical properties of a stochastic process, knowledge of the function $f(x_1, x_2, \ldots, x_n; t_1, t_2, \ldots, t_n)$ is required for every x_i, t_i and n. Most studies are limited to consideration of the 'two-point' function, $f(x_1, x_2; t_1, t_2)$.

6.1 Stationary stochastic processes

A stochastic process $x(t)$ is stationary if its pdf $P[x(t)]$ is invariant under a time shift. This definition is sometimes considered to be a very strict

44

definition of a stationary stochastic process, and it is termed by physical scientists *strict-sense stationarity*. There are, in fact, less restrictive definitions of stationary stochastic processes [131]. Examples include the following:

(a) A *wide-sense stationary stochastic process* is defined by the three conditions

$$E\{x(t)\} \equiv \mu, \tag{6.3}$$

$$E\{x(t_1)x(t_2)\} \equiv R(t_1, t_2), \tag{6.4}$$

where $R(t_1, t_2) = R(\tau)$ is a function of $\tau \equiv t_2 - t_1$, and

$$E\{x^2(t)\} = R(0). \tag{6.5}$$

Thus the variance of the process, $R(0) - \mu^2$, is time-independent.

(b) *Asymptotically stationary stochastic processes* are observed when the statistics of the random variables $x(t_1 + c), \ldots, x(t_n + c)$ does not depend on c if c is large.

(c) *Nth-order stationary stochastic processes* arise when the joint probability density

$$f(x_1, \ldots, x_n; t_1, \ldots, t_n) = f(x_1, \ldots, x_n; t_1 + c, \ldots, t_n + c) \tag{6.6}$$

holds not for every value of n, but only for $n \leq N$.

(d) *Stochastic processes stationary in an interval* are found when (6.6) holds for every t_i and $t_i + c$ in the interval considered.

At the end of this chapter, after a discussion about the time correlation properties of price changes, we discuss which definition of stationarity is more appropriate for the price changes in financial markets.

6.2 Correlation

The autocorrelation function $R(t_1, t_2)$ is sensitive to the average value of the stochastic process. For stochastic processes with average value different from zero, it is useful to consider the autocovariance,

$$C(t_1, t_2) \equiv R(t_1, t_2) - \mu(t_1)\mu(t_2). \tag{6.7}$$

For stationary processes, the autocovariance is

$$C(\tau) = R(\tau) - \mu^2. \tag{6.8}$$

The typical shape of $C(\tau)$ for positively correlated stochastic variables is a decreasing function, starting from $C(0) = \sigma^2$ and ending at $C(\tau) \simeq 0$ for large values of τ (see Fig. 6.1).

For the sake of simplicity and without loss of generality, we consider

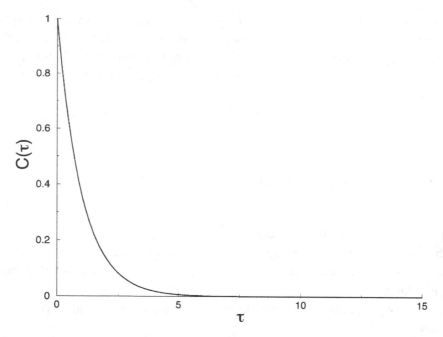

Fig. 6.1. Typical autocovariance function of a stochastic process with finite memory.

stochastic processes with zero mean and unit variance, $\mu = 0$ and $\sigma^2 = R(0) = 1$. With this choice, the autocorrelation function and the autocovariance function are the same.

Now we focus on the kind of time memory that can be observed in stochastic processes. An important question concerns the typical scale (time memory) of the autocorrelation function. For stationary processes, we can answer this important question by considering the integral of $R(\tau)$. The area below $R(\tau)$ can take on three possible values (Fig. 6.2),

$$\int_0^\infty R(\tau)d\tau = \begin{cases} \text{finite} \\ \text{infinite} \\ \text{indeterminate} \end{cases} . \tag{6.9}$$

When $\int_0^\infty R(\tau)d\tau$ is finite, there exists a typical time memory τ_c called the correlation time of the process.

Examples are the following:

Case (a): $R(\tau) = \exp[-\tau/\tau_c]$:

$$\int_0^\infty \exp\left[\frac{-\tau}{\tau_c}\right] d\tau = \tau_c. \tag{6.10}$$

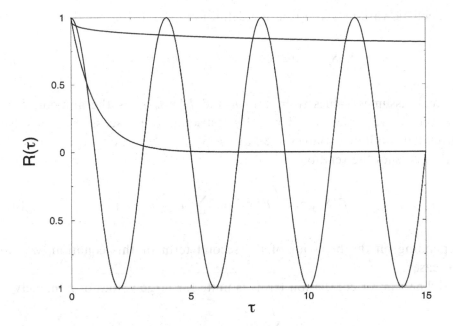

Fig. 6.2. Autocorrelation functions with and without a typical time scale.

Case (b): $R(\tau) = \exp[-\tau^{\nu}/\tau_0]$:

$$\int_0^{\infty} \exp\left[\frac{-\tau^{\nu}}{\tau_0}\right] d\tau = \frac{\tau_0^{1/\nu}}{\nu}\Gamma\left(\frac{1}{\nu}\right). \qquad (6.11)$$

Case (c): $R(\tau) \sim \tau^{\eta-1}$, where, if $0 < \eta \leq 1$,

$$\int_{t_1}^{\infty} \tau^{\eta-1} d\tau = \infty. \qquad (6.12)$$

The finiteness of the area under the autocorrelation function gives information about the typical time scale of the memory of the process. In fact, as a zero-order approximation, it is possible to model the system by saying that full correlation is present up to τ^* and no correlation is present for $\tau > \tau^*$, where τ^* is the area under the autocorrelation function. However, not all the integrals of monotonic decreasing functions are finite!

In case (c), it is impossible to select a time scale that can separate a regime of temporal correlations from a regime of pairwise independence. Random variables characterized by an autocorrelation function such as case (c) are said to be long-range correlated.

The above heuristic discussion can be formalized [27] for stationary processes by considering the general behavior of the variance of the sum S_n of

n stochastic variables x_i. From the definition (3.1), it follows that

$$E\{S_n^2\} = nE\{x_i^2\} + 2\sum_{k=1}^{n}(n-k)E\{x_i x_{i+k}\}, \qquad (6.13)$$

where k assumes values from 1 to n and $E\{x_i x_{i+k}\}$ is the autocorrelation between the variables x_i and x_{i+k}. By restricting our discussion to positively correlated random variables, we have $E\{x_i x_{i+k}\} \geq 0$. For large values of n, $E\{S_n^2\}$ satisfies the relation

$$E\{S_n^2\} \approx n\left(E\{x_i^2\} + 2\sum_{k=1}^{n}E\{x_i x_{i+k}\}\right). \qquad (6.14)$$

Depending on the behavior of the second term of this equation, we have two cases:

(i) The sum of correlation terms is finite for large values of n, namely

$$\lim_{m\to\infty}\sum_{k=1}^{n}E\{x_i x_{i+k}\} = \text{const.} \qquad (6.15)$$

In this case, it is said that the random variables are weakly dependent or *short-range correlated*. Indeed, for sufficiently large values of n, the behavior $E\{S_n^2\} = (\text{const.})n$, valid for independent random variables, still holds.

(ii) The sum of correlation terms diverges,

$$\lim_{n\to\infty}\sum_{k=1}^{n}E\{x_i x_{i+k}\} = \infty. \qquad (6.16)$$

When this condition holds, the random variables are said to be strongly dependent or *long-range correlated*. Similar random variables show a dependence on n of the variance of S_n that is stronger than linear. Long-range correlated random variables are characterized by the lack of a typical temporal scale. This behavior is observed in stochastic processes characterized by a power-law autocorrelation function as in Eq. (6.12).

We have noted that the sum S_n of n random variables can also be seen as a stochastic process in time when S_t represents a random process detected at time $t = n\Delta t$. In this case, the continuous limit of the sum of correlation terms with $n \to \infty$ is equal to $\int_{t_1}^{\infty} R(\tau)d\tau$. Hence, for time-dependent stochastic processes, the integral of the autocorrelation function can be used to distinguish between short-range correlated and long-range correlated random variables.

6.3 Short-range correlated random processes

In the previous section we noted that short-range correlated random processes are characterized by a typical time memory. One simple example is given by a stochastic process having an exponential decaying autocorrelation function (see Eq. (6.10)). This form describes, for example, the statistical memory of the velocity $v(t)$ of a Brownian particle, as the autocorrelation function of $v(t)$ is

$$R(\tau) = \sigma^2\, e^{-|\tau|/\tau_c}. \tag{6.17}$$

In addition to the characterization of the two-point statistical properties in terms of autocorrelation function, the same statistical properties might be investigated in the frequency domain. To this end, we consider the power spectrum of the random variable. The power spectrum of a wide-sense stationary random process is the Fourier transform of its autocorrelation function

$$S(f) = \int_{-\infty}^{+\infty} R(\tau)\, e^{-i2\pi f\tau} d\tau. \tag{6.18}$$

For the velocity autocorrelation function of (6.17), the power spectrum is

$$S(f) = \frac{2\sigma^2 \tau_c}{1 + (2\pi f \tau_c)^2}. \tag{6.19}$$

When $f \ll 1/(2\pi\tau_c)$, the power spectrum is essentially frequency-independent. Then, for a time window much longer than τ_c, the stochastic process is approximately white noise. The integral of white noise is called a Wiener process, a nonstationary process characterized by a power spectrum with the functional form

$$S(f) \sim \frac{1}{f^2}. \tag{6.20}$$

In summary, short-range correlated stochastic processes can be characterized with respect to their second-order statistical properties by investigating the autocorrelation function and/or the power spectrum. Fast-decaying autocorrelation functions and power spectra resembling white noise (or $1/f^2$ power spectra for the integrated variable) are 'fingerprints' of short-range correlated stochastic processes.

6.4 Long-range correlated random processes

Stochastic processes characterized by a power-law autocorrelation function (as in Eq. (6.12)) are long-range correlated. Power-law autocorrelation functions are observed in many systems – physical, biological, and economic. Let us consider a stochastic process with a power spectrum of the form

$$S(f) = \frac{\text{const.}}{|f|^{\eta}}, \tag{6.21}$$

with $0 < \eta < 2$. In the last section we saw that the case $\eta = 0$ corresponds to white noise, while $\eta = 2$ corresponds to the Wiener process. When $\eta \approx 1$, a stochastic process characterized by a spectral density as in Eq. (6.21) is called $1/f$ noise, while the general case $0 < \eta < 2$ is sometimes called $1/f^{\eta}$ noise. $1/f$ noise has been observed in a wide variety of phenomena, ranging from the current fluctuations in diodes and transistors to the fluctuations in traffic flow on a highway [46, 79, 103, 136, 156].

These stochastic processes are nonstationary. Provided that one observes the noise at times t_1 and t_2 such that the observation time T_{obs} is short compared with the time elapsed since the process began ($T_{obs} \ll t_1$), one can evaluate the autocorrelation function. Let us consider the concrete example of a noise current source with a white power spectral density driving the input of a one-dimensional resistor–capacitor transmission line of infinite length. In this system, for $0 < \eta < 1$ and $\tau > 0$, the autocorrelation function of such a nonstationary stochastic process is described by a form similar to Eq. (6.12) used for stationary processes [79],

$$R(\tau) \sim |\tau|^{\eta-1}. \tag{6.22}$$

For $1 < \eta < 2$,

$$R(t_2, \tau) \sim t_2^{\eta-1} - C|\tau|^{\eta-1}. \tag{6.23}$$

Finally, for the borderline case $\eta = 1$ and $0 < \tau \ll t_2$

$$R(t_2, \tau) \sim \ln(4t_2) - \ln|\tau|. \tag{6.24}$$

The typical shapes of these autocorrelation functions are shown in Fig. 6.3. The autocorrelation function for $1/f$ noise lacks a typical time scale, so $1/f$ noise is a long-range correlated stochastic process.

It is difficult in practice to distinguish $1/f$ noise from a process with many characteristic time scales. How many characteristic scales does one require in order to mimic a $1/f$ noise over a given frequency interval? Strictly speaking, one requires an infinite number. However, if only a finite accuracy is required, then a finite number of characteristic scales is sufficient. It has been estimated [79] that a $1/f$ power spectral density extending over 10 orders of magnitude can be mimicked at a 5 percent accuracy by the response of a linear system in which at least 8 different time scales are present, while for a 1 percent accuracy the minimal number of time scales needed is of the order of 40.

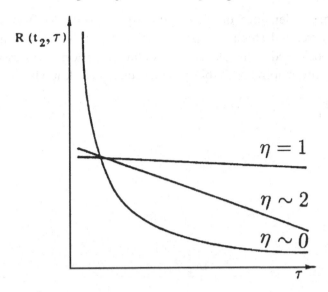

Fig. 6.3. Shapes of the autocorrelation functions for a $1/f^\eta$ noise, for the cases $\eta = 0$, 1, and 2. After Keshner [79].

6.5 Short-range compared with long-range correlated noise

If a time scale τ_c characterizes the memory of a stochastic process, then for time intervals longer than τ_c the conditional probability densities verify the equation

$$f(x_1, x_2, \ldots, x_{n-1}; t_1, t_2, \ldots, t_{n-1} | x_n; t_n) = f(x_{n-1}; t_{n-1} | x_n; t_n). \tag{6.25}$$

Stochastic processes with the above form for their conditional probability density are called Markov processes. For the simplest Markov process,

$$f(x_1, x_2, x_3; t_1, t_2, t_3) = f(x_1, t_1) f(x_1; t_1 | x_2; t_2) f(x_2; t_2 | x_3; t_3). \tag{6.26}$$

Thus only the first- and second-order conditional probability densities $f(x_1, t_1)$ and $f(x_n; t_n | x_{n+1}; t_{n+1})$ are needed to fully characterize the stochastic process. Stochastic processes lacking a typical time scale, such as $1/f$ noise, are not Markov processes.

The knowledge of the first- and second-order conditional probability densities fully characterizes a Markov process since any higher-order joint probability density can be determined from them. For a non-Markovian process, this knowledge is not sufficient to fully characterize the stochastic process.

Non-Markovian stochastic processes with the same first-order and second-order conditional probability densities are, in general, different because the joint probability densities of all orders are required to fully characterize long-range correlated stochastic processes. Thus, different $1/f$ noise signals cannot be considered to be the *same* stochastic process, unless information about higher-order joint probability densities is also known.

7

Time correlation in financial time series

In this chapter, we apply the concepts developed in Chapter 6 to discuss empirical observations of the temporal correlations detected in time series of price of financial goods. We will see that there are only short-range correlations in price changes, but there are long-range correlations in the volatility. Further, we shall discuss the degree of stationarity of financial time series.

7.1 Autocorrelation function and spectral density

Pairwise independence of the changes of the logarithm of price of a financial asset is typically investigated by analyzing the autocorrelation function of time variations of the logarithm of price (Fig. 7.1) or the spectral density of the time series of the logarithm of price itself (Fig. 7.2). These two statistical properties are equivalent for stationary stochastic processes. One finds for individual stocks that the spectral density of the logarithm of stock price is well described by the functional form (cf. Fig. 7.2)

$$S(f) \sim \frac{1}{f^2},\qquad(7.1)$$

which is the prediction for the spectral density of a random walk.

The autocorrelation function of changes of the logarithm of price is a fast-decaying function usually characterized by a correlation time much shorter than a trading day. Accurate detection of the correlation time is possible by analyzing high-frequency (intraday) data. For example, one detects a correlation time of the order of a few trading minutes by analyzing the high-frequency data of the time changes of the S&P 500 index (Fig. 7.3).

The investigation of high-frequency data allows one to extend the analysis of spectral density over a large number of frequency decades, even if the

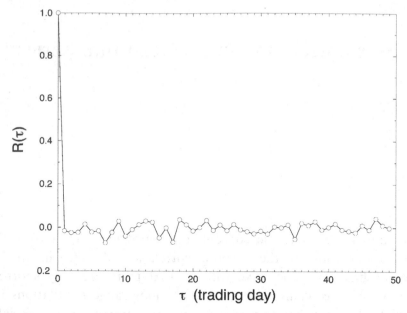

Fig. 7.1. Autocorrelation function of the logarithm of price changes for Coca Cola daily data for the period 7/89 to 10/95. The time memory is less than one trading day.

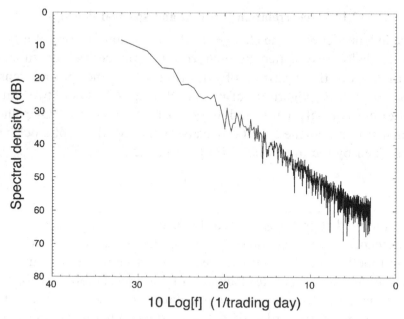

Fig. 7.2. Spectral density of the logarithm of price of Coca Cola, using daily data for the period 7/89 to 10/95. The spectral density is well approximated by a power law, $S(f) \sim 1/f^2$.

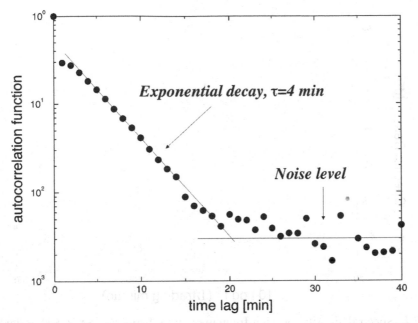

Fig. 7.3. Semi-log plot of the autocorrelation function for the S&P 500, sampled at a 1 min time scale. The straight line corresponds to exponential decay with a characteristic decay time of $\tau = 4$ min. It is apparent that after about 20 min the correlations are at the level of noise. Courtesy of P. Gopikrishnan.

total time interval over which data are analyzed is not very long. Thus high-frequency data can be useful to overcome problems associated with nonstationarity of fluctuations of economic indicators.

In Fig. 7.4 we show the spectral density of the S&P 500 index, using data recorded during the 4-year period from January 1984 to December 1987. By using high-frequency data, we can analyze the spectral density over a frequency interval of about five orders of magnitude. The data support (7.1), in agreement with the hypothesis that the stochastic dynamics of the logarithm of stock price and of a stock index may be described by a random walk.

Spectral densities and autocorrelation functions are statistical tools which are not extremely sensitive to long-range correlations. Another test, often more efficient in detecting the presence of long-range correlations, is based on the investigation of the time evolution of the standard deviation $\sigma(t)$ of price changes. In general

$$\sigma(t) \sim t^{\nu}, \tag{7.2}$$

where $\nu = 1/2$ for independent price changes. Empirical investigations of the time evolution of the standard deviation of price changes have recently

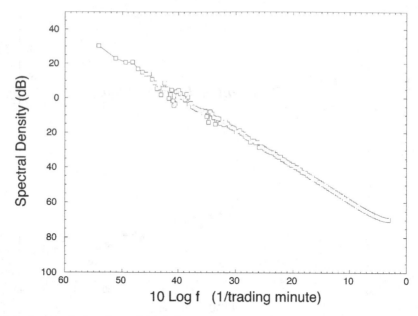

Fig. 7.4. Spectral density of high-frequency data from the S&P 500 index. The behavior of Eq. (7.1) is observed for almost five decades, with only a small deviation detected for the highest frequency investigated. Adapted from [112].

been carried out [41, 112]. The empirical behavior detected in market data is described by Eq. (7.2) with values of $v \approx 0.5$ in the time window from approximately 30 trading minutes to 100 trading days. The value of v is specific to the market investigated. Other studies analyze daily data on stock indices of New York (the New York Composite Index), Frankfurt (the DAX index), and Milan (the MIB index) exchanges, with the results for v being 0.52, 0.53, and 0.57 respectively. The values obtained show the presence of a weak long-range correlation (the empirical values of v are always slightly larger than 0.5). The strength of the long-range correlation is market-dependent and seems to be larger for less efficient markets.

Using high-frequency data for the S&P 500 index, one finds that $\sigma(t)$ has two regimes. For short times ($t < 30$ trading minutes) a superdiffusive ($v > 0.5$) behavior is observed, while in the long time regime the behavior is close to diffusive ($v = 0.5$). In the short time regime $v \approx 0.8$; this superdiffusive behavior is probably due to the fact that the time series has a memory of only a few minutes (Fig. 7.3), although it could also depend on the degree to which the process is non-Gaussian. In the long time regime covering the time interval from 30 to 10^4 trading minutes, one finds $v = 0.53$, so only a weak long-range correlation is present.

7.2 Higher-order correlations: The volatility

The autocorrelation function of price changes has exponential decay with small characteristic times – a few trading minutes for the S&P500 index (Fig. 7.3). However, pairwise independence does not directly imply that the price changes are independent random variables. Several studies performed by economists and physicists have shown that the autocorrelation function of nonlinear functions of price changes has a much longer time memory. Indeed nonlinear functions such as the absolute value or the square are long-range correlated for stock market indices and for foreign exchange currency rates.

The presence of long-range correlation in the square value of price changes suggests that there might be some other fundamental stochastic process in addition to the price change itself. This process is often referred to as volatility. The volatility is often estimated by calculating the standard deviation of the price changes in an appropriate time window. One can also use other ways of estimating it, for example by averaging the absolute values of the price changes, by maximum likelihood methods or by Bayesian methods (see [129] for a review). There are several motivations for considering the statistical properties of volatility itself. (i) Volatility can be directly related to the amount of information arriving in the market at a given time. For example, if there is large amount of information arriving in the market, then the traders would act accordingly – resulting in a large number of trades, and, in general, in large volatility. (ii) Volatility can be directly used in the modeling of the stochastic process governing the price changes, as for example in ARCH/GARCH models, to be discussed in Chapter 10. (iii) From a practical point of view, volatility is a key parameter in the measure of the risk of a financial investment.

The autocorrelation function of the volatility, estimated either as a local average of the absolute value of price changes or by the local standard deviation, is well described by a power-law decay [31, 36, 41, 95, 137]. Figure 7.5 shows the autocorrelation function for the absolute values of 1 min S&P 500 price changes using the same data as plotted in Fig. 7.3. In this case, a power-law decay with an exponent $\gamma \approx 0.3$ [96] is a good fit to the autocorrelation function.

Long-range correlations in the absolute value of price changes can also be investigated by considering the power spectrum. Figure 7.6 shows the power spectrum of absolute value of price changes of the S&P 500 index – measured in a one-hour interval. The power spectrum results are consistent with the autocorrelation function results, namely a $1/f^{\eta}$ behavior with $\eta = 1 - \gamma \approx 0.7$ [95, 96, 114].

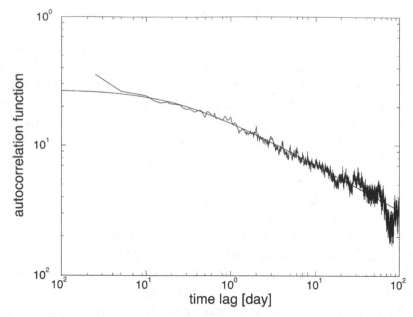

Fig. 7.5. Log–log plot of the volatility autocorrelation function using the same data as in Fig. 7.3. The solid line is a power-law regression fit over the entire range, which gives an estimate of the power-law exponent $\gamma \approx 0.3$ that quantifies the long-range correlations in the autocorrelation function. Courtesy of P. Gopikrishnan.

Studies on the distribution of volatility report a log-normal distribution for the volatility near the center of the distribution [31, 96, 133], while another work suggests that the asymptotic behavior displays power-law behavior [96]. Before concluding, we note that the existence of volatility correlation does not contradict the observation of pairwise independence of price changes because the autocorrelation of price changes depends on the second-order conditional probability density, while the volatility autocorrelation is affected by higher-order conditional probability densities.

7.3 Stationarity of price changes

From the empirical investigations discussed in previous sections, we conclude that the stochastic dynamics of price of a financial good can be approximately described by a random walk characterized by a short-range pairwise correlation. Can we describe price changes in terms of a stationary process? Empirical analyses of financial data show that price changes cannot be described by a strict-sense stationary stochastic process, since the standard deviation of price changes, namely the volatility, is time-dependent in real markets. Hence, the form of stationarity that is present in financial markets

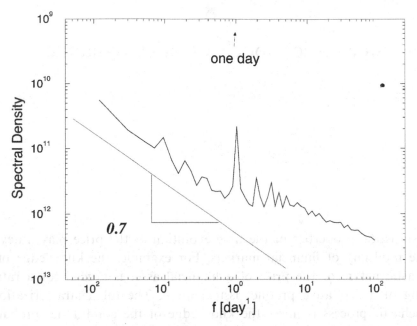

Fig. 7.6. Power spectrum of the volatility of high-frequency S&P 500 time series for the 13-year period Jan 1984 to Dec 1996. The straight line shown is not a fit to the data, but is the prediction for the power-law exponent −0.7 that is consistent with the fit to the data of Fig. 7.5. The sharp peak observed for a frequency of approximately one inverse day is related to intraday fluctuations in the volatility. Courtesy of P. Gopikrishnan.

is at best asymptotic stationarity. By analyzing a sufficiently long time series, the asymptotic pdf of prices changes is obtained. The asymptotic pdf gives the large time statistical properties of the stochastic process.

7.4 Summary

In this chapter, we have discussed several facts. (i) The statement 'price changes are pairwise uncorrelated' describes quite well the statistical behavior observed in empirical data. (ii) A short-time memory of only a few minutes is observed in the changes in financial indices. (iii) A weak long-range memory appears to be present in price changes as observed in the time evolution of $\sigma(t)$. (iv) The volatility is long-range correlated with a spectral density of the $1/f$ type.

8

Stochastic models of price dynamics

The statistical properties of the time evolution of the price play a key role in the modeling of financial markets. For example, the knowledge of the stochastic nature of the price of a financial asset is crucial for a rational pricing of a derivative product issued on it. The full characterization of a stochastic process requires the knowledge of the conditional probability densities of all orders. This is an incredible task that cannot be achieved in practice. The usual empirical approach used by physicists is performed in two steps. The first concerns the investigation of time correlation and power spectrum, while the second concerns the study of the asymptotic pdf.

The most common stochastic model of stock price dynamics assumes that $\ln Y(t)$ is a diffusive process, and the $\ln Y(t)$ increments are assumed to be Gaussian distributed. This model, known as geometric Brownian motion, provides a first approximation of the behavior observed in empirical data. However, systematic deviations from the model predictions are observed, the empirical distributions being more leptokurtic than Gaussian distributions (Fig. 8.1). A highly leptokurtic distribution is characterized by a narrower and larger maximum, and by fatter tails than in the Gaussian case. The degree of leptokurtosis is much larger for high-frequency data (Fig. 8.2).

Based on theoretical assumptions and empirical analyses, several alternative models to geometric Brownian motion have been proposed. The models differ among themselves not only with respect to the shape and leptokurtosis of the pdf, but also with respect to key properties such as

  (i) the finiteness or infiniteness of the second and higher moments of the distribution;
 (ii) the nature of stationarity present on a short time scale or asymptotically;
(iii) the continuous or discontinuous character of $Y(t)$ – or $\ln Y(t)$; and
(iv) the scaling behavior of the stochastic process.

60

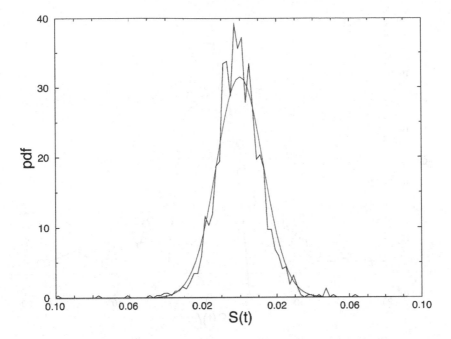

Fig. 8.1. Empirical pdf for the logarithm of daily price differences of Chevron stock traded in the New York Stock Exchange in the period 1989 to 1995. The smooth line is the Gaussian pdf with the same variance calculated from the data.

To elucidate these concepts, we first discuss a partial subset of these models, (i) the Lévy stable non-Gaussian model [102], (ii) the Student's t-distribution [19], (iii) the mixture of Gaussian distributions [32], and (iv) the truncated Lévy flight [110].

Other prominent models include the jump-diffusion model [121] and the hyperbolic-distributed stochastic process [47]. Stochastic models having a time-dependent variance over short time intervals are frequently modeled in terms of **autoregressive conditional heteroskedasticity (ARCH)** processes or, in generalized form, GARCH processes, as will be discussed in Chapter 10.

8.1 Lévy stable non-Gaussian model

The first model to take into account explicitly the leptokurtosis empirically observed in the probability density function $P(S)$ was proposed in 1963 when Mandelbrot modeled $\ln Y(t)$ for cotton prices as a stochastic process with Lévy stable non-Gaussian increments. His finding was supported by the investigations of Fama in 1965 [52], which were performed by analyzing stock

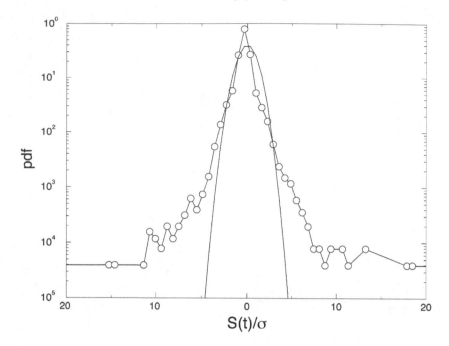

Fig. 8.2. Empirical probability density function for high-frequency price differences of the Xerox stock traded in the New York Stock Exchange during the two-year period 1994 and 1995. This semi-logarithmic plot shows the leptokurtic nature observed in empirical investigations. For comparison, the Gaussian with the measured standard deviation is also shown. Courtesy of P. Gopikrishnan.

prices in the New York Stock Exchange. The most interesting properties of Lévy stable non-Gaussian processes are

- their stability (i.e., their self-similarity), and
- their relation with a limit theorem – they are attractors in probability space.

Mandelbrot's Lévy stable hypothesis implies that $\ln Y(t)$ undergoes a discontinuous time evolution and $S(t) \equiv \ln Y(t+1) - \ln Y(t)$ is characterized by a non-Gaussian scaling and by a distribution with infinite second and higher moments. Since 1963, many papers have been devoted to considering the important problem of the finiteness or infiniteness of the variance of $S(t)$.

8.2 Student's *t*-distribution

Gaussian processes possess a finite variance. Lévy stable non-Gaussian processes possess infinite variance. Is there something 'in between' these two limits? Indeed, the Student's t-distribution is the distribution

$$P(z) = \frac{C_n}{(1 + z^2/n)^{(n+1)/2}} \tag{8.1}$$

of a stochastic process

$$z \equiv \frac{x\sqrt{n}}{\sqrt{y_1^2 + \cdots + y_n^2}}, \tag{8.2}$$

obtained from independent stochastic variables $y_1, y_2, \ldots, y_n$ and x, each with normal density, zero mean, and unit variance. Here

$$C_n \equiv \frac{\Gamma[(n+1)/2]}{\sqrt{\pi n}\,\Gamma(n/2)}. \tag{8.3}$$

When $n = 1$, $P(z)$ is the Lorentzian distribution. When $n \to \infty$, $P(z)$ is the Gaussian distribution. In general, $P(z)$ has finite moments for $k < n$. Hence a stochastic process characterized by a Student's t-distribution may have both finite and infinite moments. By varying the control parameter n (which controls the finiteness of moments of order k), one can approximate with good accuracy the log price change distribution determined from market data at a given time horizon [19].

The Student's t-distribution is, for $n \neq 1$ and finite, not stable. This implies that its shape is changing at different time horizons and that distributions at different time horizons do not obey scaling relations.

8.3 Mixture of Gaussian distributions

Another model that is capable of describing the leptokurtic behavior observed in empirical data, and that is compatible with the existence of a finite second moment of price changes, was proposed by Clark [32]. His model utilizes the concept of a subordinated stochastic process [57]. When a stochastic process occurs at times $t_1, t_2, t_3, \ldots$, which are themselves a realization of a stochastic process, starting from the random times t_i one can obtain a function $\Omega(t)$, called the directing process. Starting from the process $\ln Y(t)$ occurring at random times $t_1, t_2, t_3, \ldots$, a new random process $\ln Y[\Omega(t)]$ may therefore be formed. The process $\ln Y[\Omega(t)]$ is said to be subordinated to $\ln Y(t)$ and the distribution of the differences of the logarithm of price increments $S[\Omega(t)]$ is said to be subordinated to the distribution of $S(t)$.

From analysis of market activity, it is known that the number of transactions occurring in the market in a given time period fluctuates. Clark assumed that the trading volume is a plausible measure of the evolution of price dynamics. He used as a directing process $\Omega(t)$, the cumulative trading volume up to time t. In his model, the distribution of log price increments occurring from a given level of trading volume $P[S[\Omega(t)]]$ is subordinate to the one of individual trade $P[S(t)]$ and directed by the distribution of the trading volume $P(\Omega)$. By assuming $P[S(t)]$ to be Gaussian and $P(\Omega)$ to have all moments finite, Clark was able to prove that $P[S[\Omega(t)]]$ is a leptokurtic distribution with all the moments finite.

Clark interpreted the leptokurtic behavior observed in empirical analyses as the result of the fact that the trading activity is not uniformly distributed during the trading interval. In his model the second moment of the $P[S[\Omega(t)]]$ distribution is always finite provided $P(\Omega)$ has a finite second moment. The specific form of the distribution depends on the distribution of the directing process $\Omega(t)$. In general, the $P[S[\Omega(t)]]$ distributions do not possess scaling properties.

8.4 Truncated Lévy flight

Lévy stable non-Gaussian distributions obey scaling relations but have infinite variance. Student's t-distributions and mixtures of Gaussian distributions do not, in general, show scaling features and may or may not have finite variance. A stochastic process with finite variance and characterized by scaling relations in a large but finite interval is the truncated Lévy flight (TLF) process [110]. The TLF distribution is defined by

$$P(x) \equiv \begin{cases} 0 & x > \ell \\ cP_L(x) & -\ell \leq x \leq \ell \\ 0 & x < -\ell \end{cases}, \tag{8.4}$$

where $P_L(x)$ is the symmetric Lévy distribution of index α and scale factor γ, and c is a normalizing constant. A TLF is not a stable stochastic process, since we showed above that only Lévy distributions are stable.

Since it has a finite variance, the TLF will converge to a Gaussian process. How quickly will it converge? To answer this question, we consider the quantity $S_n \equiv \sum_{i=1}^{n} x_i$, where x_i is a truncated Lévy process, and $\langle x_i x_j \rangle = $ const δ_{ij}. The distribution $P(S_n)$ well approximates $P_L(x)$ in the limit $n \to 1$, while $P(S_n) = P_G(S_n)$ in the limit $n \to \infty$. Hence there exists a crossover value of $n, n_\times$, such that (Fig. 8.3)

$$P(S_n) \approx \begin{cases} P_L(S_n) & \text{when } n \ll n_\times \\ P_G(S_n) & \text{when } n \gg n_\times \end{cases}, \tag{8.5}$$

where $P_G(S_n)$ is a Gaussian distribution. The crossover value $n_\times$ is given by

$$n_\times \simeq A\ell^\alpha, \tag{8.6}$$

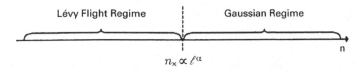

Fig. 8.3. Schematic illustration of our results for the TLF. Shown is the crossover found between Lévy flight behavior for small n and Gaussian behavior for large n. The crossover value $n_\times$ increases rapidly with the cutoff length ℓ. Adapted from [110].

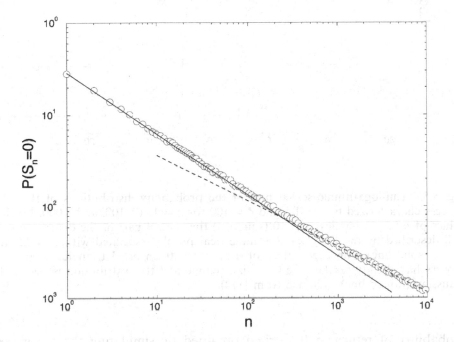

Fig. 8.4. Probability of return to the origin of S_n as a function of n for $\alpha = 1.5$ and $\ell = 100$. The simulations (circles), obtained with 5×10^4 realizations, are compared with the Lévy regime (solid line) and the asymptotic Gaussian regime calculated for $\ell = 100$ (dotted line). Adapted from [114].

where, for $\gamma = 1$,

$$A = \left[\frac{\pi\alpha}{2\Gamma(1/\alpha)[\Gamma(1+\alpha)\sin(\pi\alpha/2)/(2-\alpha)]^{1/2}} \right]^{2\alpha/(\alpha-2)}. \qquad (8.7)$$

It is possible to numerically investigate the convergence process, as n increases, of the TLF to its asymptotic Gaussian. To generate a Lévy stable stochastic process of index α and scale factor $\gamma = 1$, we use Mantegna's algorithm [106]; other algorithms exist [140]. In Fig. 8.4, we show the

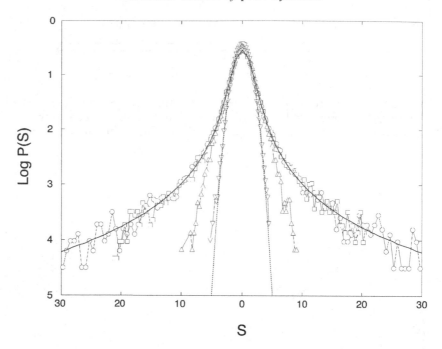

Fig. 8.5. Semi-logarithmic scaled plot of the probability distributions of the TLF process characterized by $\alpha = 1.5$ and $\ell = 100$ for $n = 1$, 10, 100, and 1,000. For low values of n ($n = 1$ (circles) and 10 (squares)) the central part of the distributions is well described by the Lévy stable symmetrical profile associated with $\alpha = 1.5$ and $\gamma = 1$ (solid line). For large values of n ($n = 1,000$ (inverted triangles)), the TLF process has already reached the Gaussian regime and the distribution is essentially Gaussian (dotted line). Adapted from [114].

probability of return to the origin obtained by simulating the S_n process when $\alpha = 1.5$ and $\ell = 100$. We also show the asymptotic behaviors for small and large n. We see clearly the crossover between the two regimes. For the selected control parameters, the crossover $n_\times$ is observed for $n_\times \approx 260$.

For the same control parameters, we also investigate the distribution $P(S_n)$ at different values of n, by simulating a TLF for $n = 1$, 10, 100, and 1,000 (Fig. 8.5). In order to be able to compare the shapes of the distribution at different values of n, we plot the distributions using the scaled variables $\tilde{P}(\tilde{S}) \equiv P(S)/n^{-1/\alpha}$, and $\tilde{S} \equiv S/n^{1/\alpha}$. From Fig. 8.5, it is clear that the TLF distribution is changing shape as a function of n. For low values of n ($n = 1$ and 10), we find good agreement with a Lévy profile, while for large values of n ($n = 1,000$), the distribution is well approximated by the asymptotic Gaussian profile. By comparing the results of Figs. 8.4 and 8.5, we note that the probability of return to the origin indicates with high accuracy the degree of convergence of the process to one of the two asymptotic regimes. For example, when $n = 1$ and 10, the probability of return is clearly in the Lévy regime (Fig. 8.4) and the central part of the TLF distribution is well described by a Lévy distribution (Fig. 8.5). Conversely, for $n = 1,000$ the probability of return to the origin is in the Gaussian regime (Fig. 8.4), and the distribution almost coincides with the Gaussian distribution characterized by the appropriate standard deviation (Fig. 8.5).

To summarize, by investigating the probability of return to the origin of an almost stable non-normal stochastic process with finite variance, one finds a clear crossover between Lévy and Gaussian regimes. Hence a Lévy-like probability distribution can be empirically observed for a long (but finite) interval of time, even in the presence of stochastic processes characterized by a *finite* variance.

In Chapter 4 we concluded that, under the efficient market hypothesis, the price change distribution for long horizons is well approximated by an infinitely divisible pdf. The TLF discussed thus far is not infinitely divisible because the truncation of the distribution is abrupt. However, an example of infinitely divisible TLFs was introduced by Koponen [86], who considered a TLF with a smooth (exponential) cutoff, and found the characteristic function

$$\varphi(q) = \exp \left\{ c_0 - c_1 \frac{(q^2 + 1/\ell^2)^{\alpha/2}}{\cos(\pi\alpha/2)} \cos[\alpha \arctan (\ell|q|)] \right\}, \qquad (8.8)$$

where c_1 is a scaling factor and

$$c_0 \equiv \frac{\ell^{-\alpha}}{\cos(\pi\alpha/2)}, \qquad (8.9)$$

A process with $\varphi(q)$ given by Eq. (8.8) is infinitely divisible since processes whose characteristic functions have an exponential form are infinitely divisible. The detailed form of the cutoff does not change the overall behavior of the convergence of the TLF to the associated asymptotic Gaussian process, since according to the Berry–Esséen theorem, the convergence is essentially controlled by the third moment of $|x|$ [144].

9

Scaling and its breakdown

No model exists for the stochastic process describing the time evolution of the logarithm of price that is accepted by all researchers. In this chapter we present one view. To this end, we discuss the results of recent empirical studies designed to answer the following questions:

 (i) Is the second moment of the price-change distribution finite?
(ii) Is self-similarity present?
(iii) If self-similarity is present, what is its nature?
(iv) Over what time interval is self-similarity present?

9.1 Empirical analysis of the S&P 500 index

We first consider a study of the statistical properties of the time evolution of the S&P 500 over the 6-year period January 1984 to December 1989 [111]. We label the time series of the index as $Y(t)$. The database has remarkably high resolution in time, with values of $Y(t)$ every minute, and sometimes every 15 seconds.

In this analysis, the time advances only during trading hours. First, we calculate the pdf $P(Z)$ of the index changes

$$Z_{\Delta t}(t) \equiv Y(t + \Delta t) - Y(t) \qquad (9.1)$$

occurring in a 1-minute interval (so $\Delta t = 1$ minute). The pdf (see Fig. 9.1) is

- almost symmetric,
- highly leptokurtic, and
- characterized by a non-Gaussian profile for small index changes.

We extract several subsets of non-overlapping price changes $Z_{\Delta t}(t)$ by varying Δt from 1 to 1,000 minutes. The number of records in each set decreases

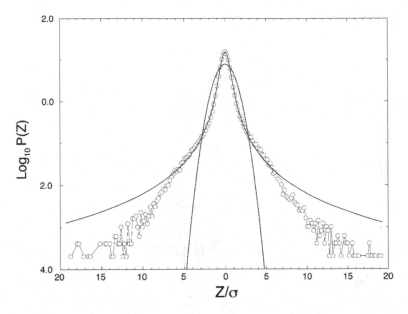

Fig. 9.1. Comparison of the $\Delta t = 1$ probability density function for high-frequency S&P 500 price changes with the Gaussian distribution (dotted line) and with a Lévy stable distribution (solid line) of index $\alpha = 1.40$ obtained from the scaling analysis and scale factor $\gamma = 0.00375$ obtained from $P(0)$ measured when $\Delta t = 1$ minute. Adapted from [111].

from 493,545 ($\Delta t = 1$ minute) to 562 ($\Delta t = 1,000$ minutes). The pdfs spread as Δt increases, as in any random process (Fig. 9.2).

When characterizing the functional form of the pdf, the usual approach is to investigate the wings. We adopt a different approach: we study the probability of return $[P_{\Delta t}(Z = 0)]$ as a function of Δt. When we plot our results on a log–log scale, we observe an interesting power-law 'scaling' behavior (Fig. 9.3). This result is compatible with a Lévy stable pdf. The index α of the Lévy distribution is the negative inverse of the slope, by Eq. (4.27). We thereby find $\alpha = 1.40 \pm 0.05$. For $\Delta t = 1$, $P_{\Delta t}(0) = 15.7$, and Eq. (4.27) results in the value $\gamma = 0.00375$ for the scale factor.

We next compare the empirical results with a Lévy stable pdf of index $\alpha = 1.40$ and scale factor $\gamma = 0.00375$. We find that there is a deviation from the Lévy distribution in the tails (Fig. 9.1). Specifically, when $|Z| \geq 6\sigma$, the data in the tails are distinctly lower than the Lévy pdf. This analysis provides an answer to question (i) by showing that the variance of price-change distribution is *finite*.

We now address question (ii). We noted already that the maxima of pdfs scale for time intervals $\Delta t \leq 1,000$ minutes. What about other regions of

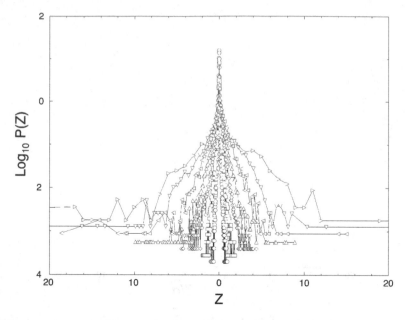

Fig. 9.2. High-frequency data for the S&P 500 index. Probability density functions of price changes measured at different time horizons $\Delta t = 1, 3, 10, 32, 100, 316, 1,000$ minutes. The typical spreading of a random walk is observed. Adapted from [111].

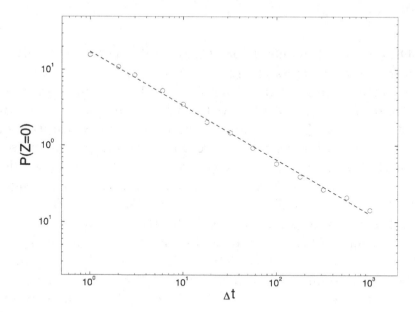

Fig. 9.3. Probability of return to the origin measured as a function of the time interval Δt. The power-law dependence is shown by plotting the measured values in a log–log plot. The slope -0.712 ± 0.025 over three orders of magnitude is consistent with a non-Gaussian scaling. Adapted from [111].

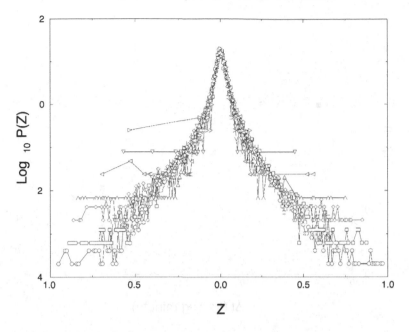

Fig. 9.4. The same probability density functions as in Fig. 9.2, but now plotted in scaled units. A good quality data collapse is observed when the scaling is performed using the value $\alpha = 1.40$. Adapted from [111].

the distribution? In Chapter 4 we observed that stable distributions are self-similar. The scaling variables for a Lévy stable process of index α are

$$\tilde{Z} \equiv \frac{Z}{(\Delta t)^{1/\alpha}} \qquad (9.2)$$

and

$$\tilde{P}(\tilde{Z}) \equiv \frac{P(Z)}{(\Delta t)^{-1/\alpha}}. \qquad (9.3)$$

When we use $\alpha = 1.40$ for the index of the Lévy distribution, the empirical results collapse well onto the $\Delta t = 1$ min distribution (Fig. 9.4).

At first glance, some of our findings seem contradictory. Specifically, we observe what at first sight appear to be inconsistent results: non-Gaussian scaling in the central part of the distribution, a Lévy non-Gaussian profile for $|Z| \leq 6\sigma$, but nevertheless a finite variance. A finite variance implies that the scaling is approximate and valid only for a finite time interval. For long time intervals, scaling must break down. To see this breakdown, we show in Fig. 9.5 $P(0)$, the probability of return to the origin measured for the S&P 500 high-frequency data, together with $P_{G}(0)$, the probability of return to the origin that would be obtained if the process were Gaussian.

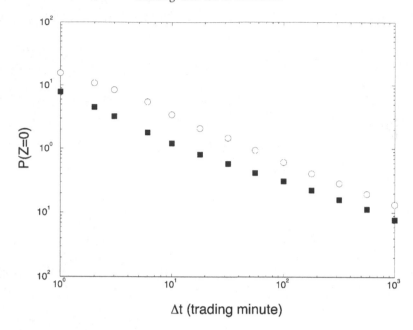

Fig. 9.5. Probability of return to the origin for S&P 500 high-frequency data (circles), together with the probability of return to the origin that would be obtained if the process were Gaussian, $P_G(0)$ (squares), estimated from the measurement of the variance for each value of Δt. The distance between the two points is a measure of the non-Gaussian nature of the pdf. Adapted from [112].

Empirical values of the variance measured at each investigated value of Δt are used to calculate $P_G(0)$. For a given value of Δt, the difference between the two probabilities of return to the origin systematically decreases for $30 < \Delta t < 1{,}000$ minutes. By extrapolating the scaling behavior of $P(0)$ and $P_G(0)$, we estimate that the breakdown of the non-Gaussian scaling occurs at approximately 10^4 trading minutes. Hence we conclude that non-Gaussian scaling is observed for a time interval that is large, but finite, ranging from 1 to approximately 10^4 trading minutes.

9.2 Comparison with the TLF distribution

Most of the empirical findings for the high-frequency changes of the S&P 500 are consistent with the simple stochastic model discussed in the previous section, the TLF. The TLF has some limitations in the modeling of empirical findings. The most important concerns the assumption of i.i.d. increments, since in a TLF model the control parameters α, γ and ℓ are time-independent. This assumption implies that the asymptotic and the short time pdfs of

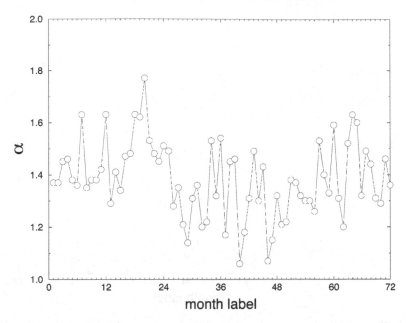

Fig. 9.6. Time dependence of the index α determined from the probability of return to the origin for the distribution of high-frequency price changes, analyzed on a monthly time scale. The determination is repeated for each of the 72 months in the period 1/84 to 12/89. Adapted from [114].

TLF increments are the same for the same time horizon Δt. We test this assumption by studying the time evolution of the α and γ parameters; the parameter γ gives one possible measure of the volatility of the process. We consider 72 subsets of the original database, and repeat for each of them the same analysis carried out on the entire database. The results are summarized in Figs. 9.6 and 9.7, where we show the time evolution of the α and γ parameters.† We conclude that α is approximately constant (Fig. 9.6) [114], while γ shows strong fluctuations, including 'bursts' of activity (Fig. 9.7). Hence empirical data show that price changes cannot be modeled in terms of a stochastic process with i.i.d. increments.

In summary, the TLF model well describes the asymptotic price-change distributions measured at different time horizons and their scaling properties, but fails to describe in a proper way the time-dependent volatility observed in market data.

---

† The ℓ parameter is not obtainable because a larger number of records is needed to estimate this parameter reliably.

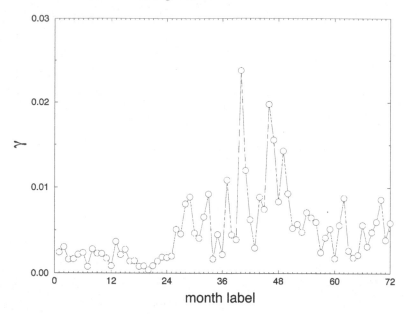

Fig. 9.7. Time dependence of the scale factor γ determined by using the value of α given in Fig. 9.6 and the probability of return to the origin of the distribution of high-frequency price changes measured for $\Delta t = 1$ minute. The determination is repeated for each of the 72 months in the period 1/84 to 12/89. Adapted from [114].

9.3 Statistical properties of rare events

One key point in the description of statistical properties of stock prices concerns 'rare events', namely the rare occurrences of large positive or negative returns. Quantitative analysis of the statistical properties of such events is difficult, and extremely large databases (or extremely long time periods) are required to reach reliable conclusions. A study [67] has considered for the two-year period January 1994 to December 1995 the high-frequency behavior of the 1,000 largest companies (by capitalization) traded in the three major US stock markets, the New York Stock Exchange (NYSE), the American Stock Exchange (AMEX), and the National Association of Securities Dealers Automated Quotation (NASDAQ). For each company, $S(t)$ was investigated, and homogeneity between the set of companies was ensured by dividing $S(t)$ by the company's volatility, measured over the investigated time period.

The behavior of rare events in the ensemble of 1,000 companies is studied by considering the cumulative distribution of the normalized variable $g(t) \equiv S(t)/\sigma_i$ where σ_i is the volatility of company i. The cumulative probability

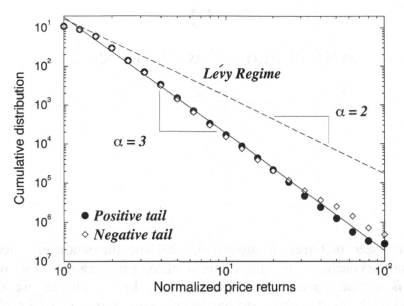

Fig. 9.8. Log–log plot of the cumulative probability distribution $F(g)$ based on high-frequency data for the 1,000 largest companies over the two-year period January 1, 1994 to December 31, 1995. The power-law behavior of Eq. (9.4) well fits the data over the range $2 \leq g \leq 100$ for both positive and negative tail. Exponents are $\alpha = 3.10 \pm 0.03$ (positive tail) and $\alpha = 2.84 \pm 0.12$ (negative tail). Adapted from [67].

distribution $F(g)$ of observing a change g or larger was found to be power-law for large values of g, both for positive and negative values of g (Fig. 9.8),

$$F(g) \sim g^{-\alpha}, \tag{9.4}$$

with exponent $\alpha \approx 3$ for both the positive and the negative tails, when the data are fit over the range $2 \leq g \leq 100$. Since $\alpha > 2$, this result is also in agreement with the conclusion that the second moment of price changes is finite [67, 68, 99].

In summary, while a definitive model for the price-change statistics does not exist, some results concerning the properties of this stochastic process have been found.

10

ARCH and GARCH processes

We have seen that there is strong empirical and theoretical evidence supporting the conclusion that the volatility of log price changes of a financial asset is a time-dependent stochastic process. In this chapter we discuss an approach for describing stochastic processes characterized by a time-dependent variance (volatility), the ARCH processes introduced by Engle in 1982 [50]. ARCH models have been applied to several different areas of economics. Examples include (i) means and variances of inflation in the UK, (ii) stock returns, (iii) interest rates, and (iv) foreign exchange rates. ARCH models are widely studied in economics and finance and the literature is huge. They can also be very attractive for describing physical systems.

ARCH models are simple models able to describe a stochastic process which is locally nonstationary but asymptotically stationary. This implies that the parameters controlling the conditional probability density function $f_t(x)$ at time t are fluctuating. However, such a 'local' time dependence does not prevent the stochastic process from having a well defined asymptotic pdf $P(x)$.

ARCH processes are empirically motivated discrete-time stochastic models for which the variance at time t depends, conditionally, on some past values of the square value of the random signal itself. ARCH processes define classes of stochastic models because each specific model is characterized by a given number of control parameters and by a specific form of the pdf, called the conditional pdf, of the process generating the random variable at time t.

In this chapter we present some widely used ARCH processes. We focus our attention on the shape of the asymptotic probability density function and on the scaling properties observed.

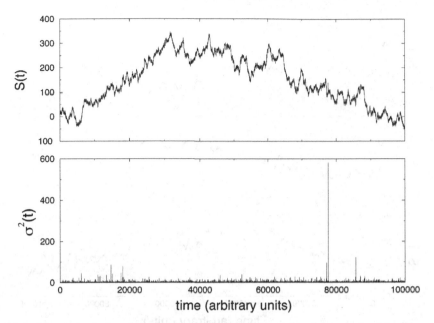

Fig. 10.1. Numerical simulation of an ARCH(1) process characterized by the parameters $\alpha_0 = 0.45, \alpha_1 = 0.55$ and conditional Gaussian probability density function. The time evolution of $S(t)$ (top) and its conditional variance (bottom) are shown.

10.1 ARCH processes

A stochastic process with autoregressive conditional heteroskedasticity, namely a stochastic process with 'nonconstant variances conditional on the past, but constant unconditional variances' [50] is an ARCH(p) process defined by the equation

$$\sigma_t^2 = \alpha_0 + \alpha_1 x_{t-1}^2 + \cdots + \alpha_p x_{t-p}^2. \tag{10.1}$$

Here $\alpha_0, \alpha_1, \ldots, \alpha_p$ are positive variables and x_t is a random variable with zero mean and variance σ_t^2, characterized by a conditional pdf $f_t(x)$. Usually $f_t(x)$ is taken to be a Gaussian pdf, but other choices are possible.

By varying the number p of terms in Eq. (10.1), one can control the amount and the nature of the memory of the variance σ_t^2. Moreover, the stochastic nature of the ARCH(p) process is also changed by changing the form of the conditional pdf $f_t(x)$. An ARCH(p) process is completely determined only when p and the shape of $f_t(x)$ are defined.

We consider the simplest ARCH process, namely the ARCH(1) process with Gaussian conditional pdf. The ARCH(1) process is defined by

$$\sigma_t^2 = \alpha_0 + \alpha_1 x_{t-1}^2, \tag{10.2}$$

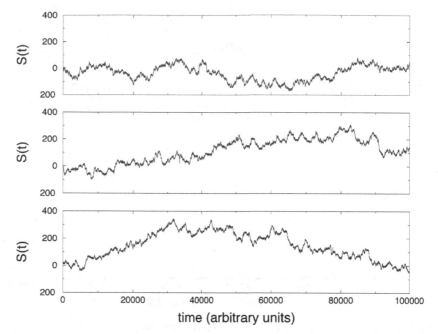

Fig. 10.2. Numerical simulations of ARCH(1) processes with the same unconditional variance ($\sigma^2 = 1$) and different values of the unconditional kurtosis. Top: $\alpha_0 = 1$, $\alpha_1 = 0$ (so $\kappa = 3$ by Eq. (10.6)). Middle: $\alpha_0 = \alpha_1 = 0.5$ (so $\kappa = 9$). Bottom: $\alpha_0 = 0.45$, $\alpha_1 = 0.55$ (so $\kappa = 23$).

and

$$S(t) = \sum_{i=1}^{i=t} x_i. \tag{10.3}$$

In Fig. 10.1 we show the time evolution of $S(t)$ obtained by simulating an ARCH(1) process with parameters $\alpha_0 = 0.45$ and $\alpha_1 = 0.55$. In the same figure we also show the time evolution of the variance σ_t^2. Although the conditional pdf is chosen to be Gaussian, the asymptotic pdf presents a given degree of leptokurtosis because the variance σ_t of the conditional pdf is itself a fluctuating random process.

An ARCH(1) process with Gaussian conditional pdf is characterized by a finite 'unconditional' variance (the variance observed on a long time interval), provided

$$1 - \alpha_1 \neq 0 \qquad 0 \leq \alpha_1 < 1. \tag{10.4}$$

The value of the variance is

$$\sigma^2 = \frac{\alpha_0}{1 - \alpha_1}. \tag{10.5}$$

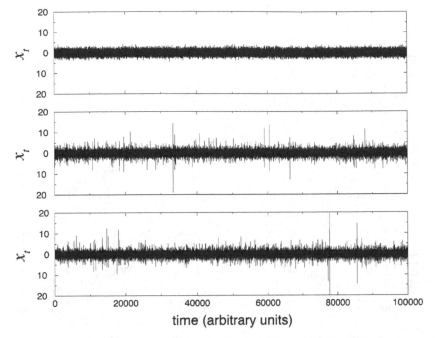

Fig. 10.3. Successive increments of the simulations shown in Fig. 10.2. Events outside three standard deviations are almost absent when $\kappa = 3$ (top), are present when $\kappa = 9$ (middle), and are more intense when $\kappa = 23$ (bottom).

The kurtosis of the ARCH(1) process is [50]

$$\kappa \equiv \frac{\langle x^4 \rangle}{\langle x^2 \rangle^2} = 3 + \frac{6\alpha_1^2}{1 - 3\alpha_1^2}, \qquad (10.6)$$

which is finite if

$$0 \leq \alpha_1 < \frac{1}{\sqrt{3}}. \qquad (10.7)$$

Hence, by varying α_0 and α_1, it is possible to obtain stochastic processes with the same unconditional variance but with different values of the kurtosis.

Next we consider three examples of ARCH(1) time series having the same unconditional variance but different values of the kurtosis; $\sigma^2 = 1$ for all examples, while the kurtosis κ increases from 3 (Wiener process) to 23. In Fig. 10.2 we show the $S(t)$ time series, while in Fig. 10.3 we show the x time series. By inspecting Fig. 10.2 we note that the 'territory visited' in the ARCH(1) process increases for large κ (despite the fact that $\sigma^2 = 1$ for all three examples); corresponding to this observation, we see in Fig. 10.3 that jumps of size larger than 3 times the unconditional variance are observed when $\kappa > 3$. From the shape of the asymptotic pdfs (Fig. 10.4), we note the

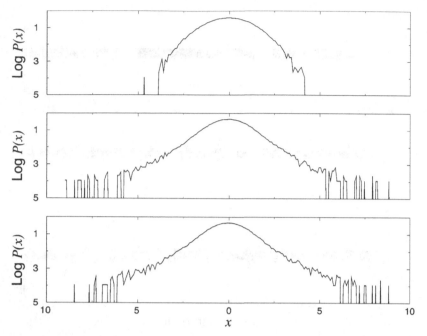

Fig. 10.4. Probability density function of the successive increments shown in Fig. 10.3. The pdf is Gaussian when $\kappa = 3$ (top) and is leptokurtic when $\kappa = 9$ or 23 (middle and bottom).

higher degree of leptokurtosis when $\kappa > 3$. When $\alpha_0 = 1$ and $\alpha_1 = 0$, the unconditional pdf $P(x)$ is Gaussian. For $0 < \alpha_1 < 1$, the exact shape of the ARCH(1) pdf is unknown.

10.2 GARCH processes

In many applications using the linear ARCH(p) model, a large value of p is required. This usually poses some problems in the optimal determination of the $p + 1$ parameters $\alpha_0, \alpha_1, \ldots, \alpha_p$, which best describe the time evolution of a given economic time series. The overcoming of this difficulty leads to the introduction of generalized ARCH processes, called GARCH(p, q) processes, introduced by Bollerslev in 1986 [20].

This class of stochastic processes is defined by the relation

$$\sigma_t^2 = \alpha_0 + \alpha_1 x_{t-1}^2 + \cdots + \alpha_q x_{t-q}^2 + \beta_1 \sigma_{t-1}^2 + \cdots + \beta_p \sigma_{t-p}^2, \qquad (10.8)$$

where $\alpha_0, \alpha_1, \ldots, \alpha_q, \beta_1, \ldots, \beta_p$ are control parameters. Here x_t is a random variable with zero mean and variance σ_t^2, and is characterized by a conditional pdf $f_t(x)$, which is arbitrary but is often chosen to be Gaussian.

We consider the simplest GARCH process, namely the GARCH(1,1) process, with Gaussian conditional pdf. It can be shown [9] that

$$\sigma^2 = \frac{\alpha_0}{1 - \alpha_1 - \beta_1},$$ (10.9)

and the kurtosis is given by the relation

$$\kappa = 3 + \frac{6\alpha_1^2}{1 - 3\alpha_1^2 - 2\alpha_1\beta_1 - \beta_1^2}.$$ (10.10)

10.3 Statistical properties of ARCH/GARCH processes

For the sake of simplicity, in this section we present the statistical properties of the GARCH(1,1) process with Gaussian conditional pdf. The properties of the more general GARCH(p, q) processes with Gaussian conditional pdf are essentially the same [20].

First we discuss the class of stochastic processes to which GARCH(1,1) belongs. The GARCH(1,1) process is defined by

$$\sigma_t^2 = \alpha_0 + \alpha_1 x_{t-1}^2 + \beta_1 \sigma_{t-1}^2.$$ (10.11)

The random variable x_t can be written in term of σ_t by defining

$$x_t \equiv \eta_t \sigma_t,$$ (10.12)

where η_t is an i.i.d. random process with zero mean, and unit variance. Under the assumption of Gaussian conditional pdf, η_t is Gaussian. By using Eq. (10.12), one can rewrite Eq. (10.11) as

$$\sigma_t^2 = \alpha_0 + (\alpha_1 \eta_{t-1}^2 + \beta_1)\sigma_{t-1}^2.$$ (10.13)

Equation (10.13) shows that GARCH(1,1) and, more generally, GARCH(p, q) processes are essentially random multiplicative processes. The autocorrelation function of the random variable x_t, $R(\tau) = \langle x_t x_{t+\tau} \rangle$ is proportional to a delta function $\delta(\tau)$.

What about the higher-order correlation of the process? Following Bollerslev [20], we will see that in a GARCH(1,1) process, x_t^2 is a Markovian random variable characterized by the time scale $\tau = |\ln(\alpha_1 + \beta_1)|^{-1}$. Hence a GARCH(1,1) process provides an interesting example of a stochastic process x_t that is second-order uncorrelated, but is higher-order correlated.

Let us first recall that a GARCH(1,1) process may be written as

$$x_t^2 = \alpha_0 + \alpha_1 x_{t-1}^2 + \beta_1 x_{t-1}^2 - \beta_1 v_{t-1} + v_t,$$ (10.14)

where

$$v_t = x_t^2 - \sigma_t^2 = (\eta_t^2 - 1)\sigma_t^2. \tag{10.15}$$

It is worth noting that v_t is serially uncorrelated with zero mean. This form of writing the GARCH(1,1) process shows that the GARCH(1,1) process can be interpreted as an autoregressive moving average (ARMA) process in x_t^2. This formulation is also useful in determining the autocovariance of x_t^2, which is defined as

$$\mathrm{cov}(x_t^2, x_{t+n}^2) \equiv \langle x_t^2 x_{t+n}^2 \rangle - \langle x_t^2 \rangle \langle x_{t+n}^2 \rangle. \tag{10.16}$$

For a GARCH(1,1) process defined as in Eq. (10.11) with a finite fourth-order moment, by using Eq. (10.14) and Eq. (10.15) it is possible to conclude that

$$\mathrm{cov}(x_t^2, x_{t+n+1}^2) = (\alpha_1 + \beta_1)\mathrm{cov}(x_t^2, x_{t+n}^2). \tag{10.17}$$

For the most general case of a GARCH(p,q) process, it is also possible to write down the relation between the autocovariance of x_t^2 and x_{t+n}^2 (time lag of n steps) with the autocovariance of x_t^2 and x_{t+n-i}^2 (time lag of $n-i$ steps). This general relation is [20]

$$\mathrm{cov}(x_t^2, x_{t+n}^2) = \sum_{i=1}^{m} B_i \mathrm{cov}(x_t^2, x_{t+n-i}^2). \tag{10.18}$$

where $m = \max\{p, q\}$ and

$$B_i = \alpha_i + \beta_i. \tag{10.19}$$

From Eq. (10.17), we see that the autocovariance of the square of the process x_t is described by the exponential form

$$\mathrm{cov}(x_t^2, x_{t+n}^2) = A e^{-n/\tau}, \tag{10.20}$$

where $A \equiv \alpha_0/(1 - B)$ and $\tau \equiv |\ln B|^{-1}$, and $B \equiv \alpha_1 + \beta_1$. In a GARCH(1,1) process the square of the process x_t^2 is a Markovian process characterized by the time scale τ.

The Markovian character of x_t^2 is also observed in ARCH processes. For example, the characteristic time scale of the autocovariance of x_t^2 is $\tau = |\ln \alpha_1|^{-1}$ in the ARCH(1) process. A difference in the temporal memory of ARCH(1) and GARCH(1,1) processes is detected by comparing the characteristic time scale for these two processes. Let us consider ARCH(1) and GARCH(1,1) processes with finite second and fourth moments. The requirement of the finiteness of the fourth moment implies that α_1 must be lower than $1/\sqrt{3}$ (see Eq. (10.7)) for the ARCH(1) process whereas the corresponding GARCH(1,1) process with finite fourth moment can be characterized by values of α_1 and β_1, such as $\alpha_1 + \beta_1$ is close to 1 provided that β_1 is larger than approximately 0.7 (the conditions for finiteness or infiniteness of moments for a GARCH(1,1) process can be found in Ref. [20]). Hence an ARCH(1) process with finite fourth moment may be characterized by a maximal characteristic time scale in the square of fluctuations of approximately $\tau = |\ln 1/\sqrt{3}|^{-1} \simeq 1.8$ time units whereas in the GARCH(1,1) process with finite fourth moment we can observe a characteristic time scale longer than hundreds of time units, the only condition being that the β_1 parameter must be larger than 0.7.

In previous chapters, we have shown that there is empirical evidence that the variance of returns is characterized by a power-law correlation. Since the correlation of the square of a GARCH(1, 1) process is exponential, a GARCH(1, 1) process cannot be used to describe this empirically observed phenomenon properly. In spite of this limitation, GARCH(1, 1) processes are widely used to model financial time series. The limitation is overcome by using values of B close to one in empirical analysis [1]. Values of B close to one imply a time memory that could be of the order of months. The model's values for the α_1 and β_1 parameters – obtained in the period 1963 to 1986 by analyzing the daily data of stock prices of the Center for Research in Security Prices (CRSP) – give $\alpha_1 = 0.07906$ and $\beta_1 = 0.90501$ [1]. The sum $B \equiv \alpha_1 + \beta_1$ is then 0.98407, which implies a memory of x_t^2 corresponding to $\tau = 62.3$ trading days. Such a long time memory in the square of returns mimics in an approximate way the power-law correlation of this variable in a finite time window.

Another key aspect of the statistical properties of the GARCH(1,1) process is its behavior for different time horizons. For finite variance GARCH(1,1) processes, the central limit theorem applies and one expects that the temporal aggregation of a GARCH(1,1) process progressively implies a decrease in the leptokurtosis of the process. Drost and Nijman [43] carried out a quantitative study of this problem. They were able to show that a 'temporal aggregation' of a GARCH(1,1) process is still a GARCH(1,1) process, but it is characterized by different control parameters. Specifically, when a GARCH(1,1) x_t is 'aggregated' as

$$S_t^{(m)} = \sum_{i=0}^{m-1} x_{t-i}. \qquad (10.21)$$

It can be shown that $S_t^{(m)}$ is also a GARCH(1,1) process characterized by

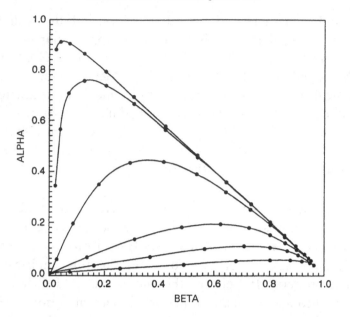

Fig. 10.5. Aggregation of GARCH(1,1). Marks indicate the parameters α_1 and β_1 of a GARCH(1,1) model generated by doubling or halving the sampling interval. The starting GARCH(1,1) processes are characterized by $\beta_1 = 0.8$ and $\alpha_1 = 0.05$, 0.1, 0.15, 0.19, 0.199, and 0.1999 (from bottom to top, respectively). After Drost and Nijman [43].

the control parameters [43]

$$\alpha_0^{(m)} = \alpha_0 \frac{1 - B^m}{1 - B},$$
$$\alpha_1^{(m)} = B^m - \beta^{(m)}$$

(10.22)

where $\beta^{(m)} \in (0, 1)$ is the solution of the quadratic equation

$$\frac{\beta^{(m)}}{1 + [\beta^{(m)}]^2} = \frac{\beta_1 B^{m-1}}{1 + \alpha_1^2 [1 - B^{2m-2}]/[1 - B^2] + \beta_1^2 B^{2m-2}}.$$

(10.23)

In Fig. 10.5 we show the behavior of the parameters $\alpha_1^{(m)}$ and $\beta_1^{(m)}$ for the temporal aggregation of GARCH(1,1) processes obtained by repeatedly doubling or halving the time interval for GARCH(1,1) processes, for a range of parameter values. When the time interval is doubled, the parameters move to lower values of $\beta^{(m)}$ while $\alpha_1^{(m)}$ may increase or decrease, depending on the starting values of α_1 and β_1. However, in any case (see the left region of Fig. 10.5), the attractor for all the GARCH(1,1) processes with finite variance is the process characterized by $\alpha_1^{(m)} = 0$, $\beta^{(m)} = 0$ – namely a Gaussian process.

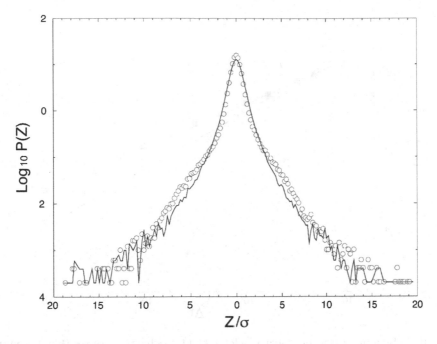

Fig. 10.6. Comparison of the empirical pdf measured from high-frequency S&P 500 data with $\Delta t = 1$ minute with the unconditional pdf of a GARCH (1,1) process characterized by $\alpha_o = 2.30 \times 10^{-5}, \alpha_1 = 0.09105$, and $\beta_1 = 0.9$ (Gaussian conditional probability density). The agreement is good for more than four decades.

In summary, for any GARCH(1,1) process, temporal aggregation implies that the unconditional pdf of the process presents a degree of leptokurtosis that decreases when the time horizon between the variables increases. Unfortunately, the knowledge of the behavior of $\alpha_1^{(m)}$ and $\beta^{(m)}$ for any value of m is not sufficient to determine the behavior of the probability of return to the origin of a GARCH(1,1) process. We investigate this function numerically in the next section, where we compare empirical findings and GARCH(1,1) simulations.

10.4 The GARCH(1,1) and empirical observations

In this section we compare empirical investigations of the S&P 500 high-frequency data with simulations of a GARCH(1,1) process. Specifically we compare the pdf and the scaling properties of our empirical analysis with the pdf and the scaling properties of a GARCH(1,1) process characterized by the same variance and kurtosis measured in the time series of the S&P 500.

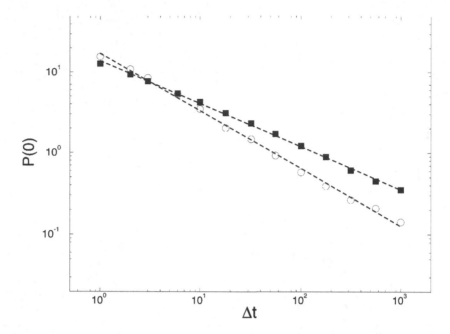

Fig. 10.7. Scaling properties of a GARCH(1,1) stochastic process (black squares), with the same control parameters as in Fig. 10.6. The scaling of the GARCH(1,1) process fails to describe the empirical behavior observed in the S&P 500 high-frequency data (which are also shown for comparison as white circles). Note that the slope 0.53 is extremely close to the Gaussian value of 0.5, indicating that the scaling is close to the scaling of a Gaussian process.

The GARCH(1,1) process has three control parameters, α_0, α_1 and β_1. We select the parameters that best describe the data by ensuring that the variance and the kurtosis of the GARCH(1,1) process equals the measured values. In this way, we determine the values of the two parameters α_0 and α_1. The value of the third parameter, β_1, is chosen to be 0.9, because this value is often used in the literature [1].

From the empirical analysis of the S&P 500 high frequency data, we find for $\Delta t = 1$ minute that $\sigma^2 = 0.00257$ and $\kappa \approx 43$. Using Eqs. (10.9) and (10.10) we obtain $\alpha_0 = 2.30 \times 10^{-5}$ and $\alpha_1 = 0.09105$.

By properly choosing the control parameters α_0, α_1, and β_1, GARCH(1,1) stochastic processes with Gaussian conditional pdfs model quite well the short-time leptokurtic pdf of price changes. In Fig. 10.6, we show the price change distribution of the S&P 500 together with the distribution observed for the GARCH(1,1) process. The agreement is quite good.

The fact that the GARCH(1,1) process describes well the $\Delta t = 1$ minute pdf does not ensure that the same process describes well the stochastic dynamics of the empirical data for any time horizon Δt. Hence an important question to be answered is whether the overall dynamics of high-frequency price changes are well described by a GARCH(1,1) process. To describe the dynamics of price changes in a complete way, in addition to pdf of price changes at a given time horizon, the scaling properties of price change pdfs need to be considered also. What about the scaling properties of GARCH(1,1) stochastic processes? A theoretical answer at the moment does not exist; however, indications can be obtained by performing numerical simulations of GARCH processes.

Figure 10.7 shows the scaling property of the probability of return to the origin for a GARCH(1,1) process with conditional Gaussian pdf for the same control parameters as those used to obtain the pdf of Fig. 10.6. The empirical behavior observed in the S&P 500 high-frequency data is also shown for comparison. Although the GARCH(1,1) process is able to describe the $\Delta t = 1$ minute pdf, it fails to describe the scaling properties of pdfs for all time horizons using the same control parameters. Thus to test the effectiveness of a model, it is not sufficient to compare distributions at a single time horizon.

10.5 Summary

ARCH and GARCH processes are extremely interesting classes of stochastic processes. They are widely used in finance, and may soon be used in other disciplines. Concerning high-frequency stock market data, ARCH/GARCH processes with Gaussian conditional pdf are able to describe the pdf of price changes at a given time horizon, but fail to describe properly the scaling properties of pdfs at different time horizons.

Open questions concerning this class of stochastic processes include:

(i) What is the form of the asymptotic pdf of the ARCH and GARCH processes characterized by a given conditional probability density function $f_t(x)$?

(ii) What is the nature of the scaling property of the probability of return to the origin as a function of the values of the control parameters and of the shape of the conditional probability density function?

11

Financial markets and turbulence

One of the objections often leveled at the approach of physicists working with economic systems is that this kind of activity cannot be a branch of physics because the 'equation of motion of the process' is unknown. But if this criterion – requiring that the Hamiltonian of the process be known or obtainable – were to be applied across the board, several fruitful current research fields in physics would be disqualified, e.g., the modeling of friction and many studies in the area of granular matter. Moreover, a number of problems in physics that are described by a well defined equation – such as turbulence [61] – are not analytically solvable, even with sophisticated mathematical and physical tools.

On a qualitative level, turbulence and financial markets are attractively similar. For example, in turbulence, one injects energy at a large scale by, e.g., stirring a bucket of water, and then one observes the manner in which the energy is transferred to successively smaller scales. In financial systems 'information' can be injected into the system on a large scale and the reaction to this information is transferred to smaller scales – down to individual investors. Indeed, the word 'turbulent' has come into common parlance since price fluctuations in finance qualitatively resemble velocity fluctuations in turbulence. Is this qualitative parallel useful on a quantitative level, such that our understanding of turbulence might be relevant to understanding price fluctuations?

In this chapter, we will discuss fully developed turbulence in parallel with the stochastic modeling of stock prices. Our aim is to show that cross-fertilization between the two disciplines might be useful, not that the turbulence analogy is quantitatively correct. We shall find that the formal correspondence between turbulence and financial systems is not supported by quantitative calculations.

11.1 Turbulence

Turbulence is a well defined but unsolved physical problem which is today one of the great challenges in physics. Among the approaches that have been tried are analytical approaches, scaling arguments based on dimensional analysis, statistical modeling, and numerical simulations.

Consider a simple system that exhibits turbulence, a fluid of kinematic viscosity v flowing with velocity V in a pipe of diameter L. The control parameter whose value determines the 'complexity' of this flowing fluid is the Reynolds number,

$$\text{Re} \equiv \frac{LV}{v}. \tag{11.1}$$

When Re reaches a particular threshold value, the 'complexities of the fluid explode' as it suddenly becomes turbulent.

The equations describing the time evolution of an incompressible fluid have been known since Navier's work was published in 1823 [128], which led to what are now called the Navier–Stokes equations,

$$\frac{\partial}{\partial t}\mathbf{V}(\mathbf{r},t) + (\mathbf{V}(\mathbf{r},t)\cdot\nabla)\mathbf{V}(\mathbf{r},t) = -\nabla P + v\nabla^2\mathbf{V}(\mathbf{r},t), \tag{11.2}$$

and

$$\nabla\cdot\mathbf{V}(\mathbf{r},t) = 0. \tag{11.3}$$

Here $\mathbf{V}(\mathbf{r},t)$ is the velocity vector at position $\mathbf{r}$ and time t, and P is the pressure. The Navier–Stokes equations characterize completely 'fully developed turbulence', a technical term indicating turbulence at a high Reynolds number. The analytical solution of (11.2) and (11.3) has proved impossible, and even numerical solutions are impossible for very large values of Re.

In 1941, a breakthrough in the description of fully developed turbulence was achieved by Kolmogorov [82–84]. He showed that in the limit of infinite Reynolds numbers, the mean square velocity increment

$$\langle [\Delta V(\ell)]^2 \rangle = \langle [V(r+\ell) - V(r)]^2 \rangle \tag{11.4}$$

behaves approximately as

$$\langle [\Delta V(\ell)]^2 \rangle \sim \ell^{2/3} \tag{11.5}$$

in the inertial range, where the dimensions are smaller than the overall dimension within which the fluid's turbulent behavior occurs and larger than the typical length below which kinetic energy is dissipated into heat.

Kolmogorov's theory describes well the second-order $\langle [\Delta V(\ell)]^2 \rangle$ and provides the exact relation for the third-order $\langle [\Delta V(\ell)]^3 \rangle$ moments observed in experiments, but fails to describe higher moments.

In fully developed turbulence, velocity fluctuations are characterized by an intermittent behavior, which is reflected in the leptokurtic nature of the pdf of velocity increments. Kolmogorov theory is not able to describe the intermittent behavior of velocity increments. In the experimental studies of fully developed turbulence, experimentalists usually measure the velocity $V(t)$ as a function of time. From this time series, the spatial dependence of the velocity $V(\ell)$ can be obtained by making the Taylor hypothesis [124].

11.2 Parallel analysis of price dynamics and fluid velocity

Turbulence displays both analogies with and differences from the time evolution of prices in a financial market. To see this, we discuss the results of a parallel analysis [112] of two systems, the time evolution of the S&P 500 index and the velocity of a turbulent fluid at high Reynolds number. Both processes display intermittency and non-Gaussian features at short time intervals. Both processes are nonstationary on short time scales, but are asymptotically stationary. A better understanding and modeling of stochastic processes that are only asymptotically stationary is of potential utility to both fields.

Specifically, we consider the statistical properties of (i) the S&P 500 high-frequency time series recorded during the six-year period 1984 to 1989 and (ii) the wind velocity recorded in the atmospheric surface layer about 6 m above a wheat canopy in the Connecticut Agricultural Research Station.† Similarities and differences are already apparent by direct inspection of the time evolutions of the index and the velocity of the fluid, as well as the successive measurements of both time series.

First, we compare the time evolution of the S&P 500 index (Fig. 11.1a) and the time evolution of fluid velocity (Fig. 11.2a). We also display one-hour changes in the S&P 500 index (Fig. 11.1b) and fluid velocity changes at the highest sampling rate (Fig. 11.2b). By analyzing the temporal evolution of successive increments in both signals, we can obtain useful information concerning the statistical properties·of the two signals. A quantitative analysis can be performed by considering the volatility for financial data, and the square root of the second moment of velocity fluctuations for turbulence data.

---

† K. R. Sreenivasan kindly provided the data on fully developed turbulence.

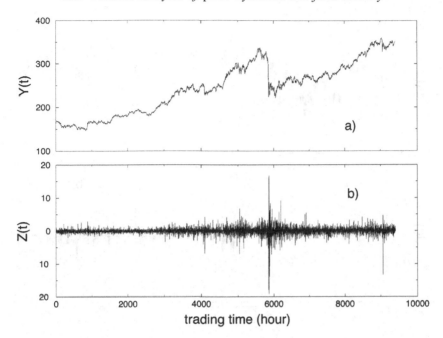

Fig. 11.1. (a) Time evolution of the S&P 500, sampled with a time resolution $\Delta t = 1$ h, over the period January 1984 to December 1989. (b) Hourly variations of the S&P 500 index in the 6-year period January 1984 to December 1989.

Both sets of data are seen in Fig. 11.3 to be well described by power laws,

$$\sigma(\Delta t) \sim (\Delta t)^{\nu}, \tag{11.6}$$

but with quite different values of the exponent ν. Index changes are essentially uncorrelated (the observed value of $\nu = 0.53$ is extremely close to $1/2$, the value expected for uncorrelated changes), while velocity changes are anti-correlated ($\nu = 0.33 < 1/2$). Thus the quantitative difference between the two forms of behavior implies that the nature of the time correlation between the successive changes must be different for the two processes. Indeed, the time evolutions of the index and the velocity in Figs. 11.1a and 11.2a look quite different, since there is a high degree of anticorrelation in the velocity. This difference is also visually apparent, from Fig. 11.2b, which is approximately symmetric about the abscissa, whereas Fig. 11.1b is not.

This difference between the two stochastic processes is also observable in the power spectra of the index and velocity time series (see Fig. 11.4). While both obey Eq. (6.21) over several frequency decades, the exponents η are quite different. For the S&P 500 index, $\eta = 1.98$, so the spectral density is essentially identical to the power spectrum of an uncorrelated random

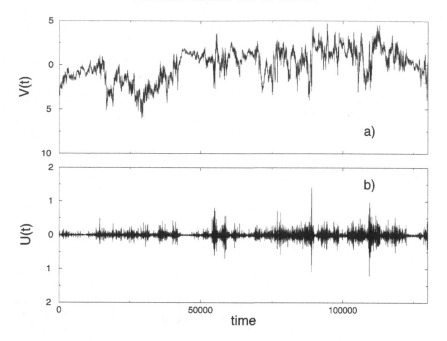

Fig. 11.2. Time evolution of the fluid velocity in fully developed turbulence. (a) Time evolution of the wind velocity recorded in the atmosphere at extremely high Reynolds number; the Taylor microscale Reynolds number is of the order of 1,500. The time units are given in arbitrary units. (b) Velocity differences of the time series given in (a). Adapted from [113].

process ($\eta = 2$). For the velocity time series, $\eta \approx 5/3$ in the inertial range and $\eta \approx 2$ in the dissipative range.

Ghashghaie *et al.* [64] have proposed a formal analogy between the velocity of a turbulent fluid and the currency exchange rate in the foreign exchange market. They supported their conclusion by observing that when measurements are made at different time horizons Δt, the shapes of the pdf of price increments in the foreign exchange market and the pdf of velocity increments in fully developed turbulence both change. Specifically, the shapes of both pdfs display leptokurtic profiles at short time horizons. However, the parallel analysis of the two phenomena [112, 113] shows that the time correlation is completely different in the two systems (Fig. 11.4). Moreover, stochastic processes such as the TLF and the GARCH(1,1) processes also describe a temporal evolution of the pdf of the increments which evolves from a leptokurtic to a Gaussian shape, so such behavior is not specific to the velocity fluctuations of a fully turbulent fluid.

To detect the degree of similarity between velocity fluctuations and index

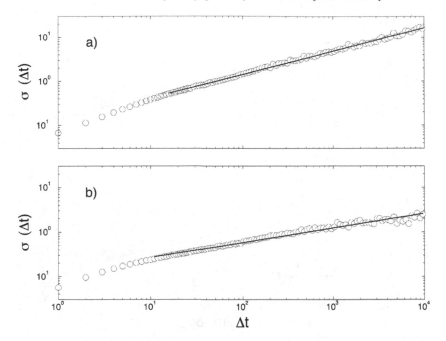

Fig. 11.3. (a) Standard deviation $\sigma(\Delta t)$ of the probability distribution $P(Z)$ characterizing the increments $Z_{\Delta t}(t)$ plotted double logarithmically as a function of Δt for
the S&P 500 time series. After a time interval of superdiffusive behavior ($0 < \Delta t \le 15$
minutes), a diffusive behavior close to the one expected for a random process with
uncorrelated increments is observed; the measured diffusion exponent 0.53 (the
slope of the solid line) is close to the theoretical value $1/2$. (b) Standard deviation
$\sigma(\Delta t)$ of the probability distribution $P(U)$ characterizing the velocity increments
$U_{\Delta t}(t) \equiv V(t + \Delta t) - V(t)$ plotted double logarithmically as a function of Δt for the
velocity difference time series in turbulence. After a time interval of superdiffusive
behavior ($0 < \Delta t \le 10$), a subdiffusive behavior close to the one expected for a fluid
in the inertial range is observed. In fact, the measured diffusion exponent 0.33 (the
slope of the solid line) is close to the theoretical value $1/3$. Adapted from [112].

changes, consider the probability of return to the origin, $P_{\Delta t}(U = 0)$, as a
function of Δt for a turbulent fluid, obtained by following the same procedure
used to obtain Fig. 9.3. We display in Fig. 11.5 the measured $P_{\Delta t}(U = 0)$.
We also show the estimated $P_G(U = 0)$ obtained starting from the measured
variance of velocity changes $\sigma(\Delta t)$ and assuming a Gaussian shape for the
distribution ($P_G(U = 0) = 1/\sqrt{2\pi}\sigma(\Delta t)$). The difference between each pair of
points is a measure of the ratio $P_{\Delta t}/P_G$, and quantifies the degree of non-
Gaussian behavior of the velocity differences. We note that the turbulence
process becomes increasingly Gaussian as the time interval Δt increases, but
we do not observe any scaling regime.

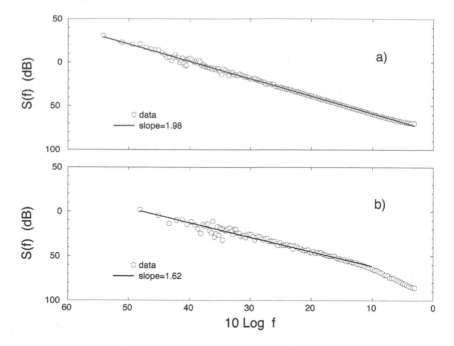

Fig. 11.4. (a) Spectral density of the S&P 500 time series. The $1/f^2$ power-law behavior expected for a random process with increments that are pairwise independent is observed over a frequency interval of more than four orders of magnitude. (b) Spectral density of the velocity time series. The $1/f^{5/3}$ inertial range (low frequency) and the dissipative range (high frequency) are clearly observed. Adapted from [112].

11.3 Scaling in turbulence and in financial markets

The concept of scaling is used in a number of areas in science, emerging when an investigated process does not exhibit a typical scale. A power-law behavior in the variance of velocity measurements in turbulence (see Eq. (11.5)) is an example of a scaling behavior, as is the power-law behavior of volatility at different time horizons in financial markets (see Eq. (11.6)). The reasons underlying the two scaling behaviors are, however, quite different. In the turbulence case, the 2/3 exponent of the distance ℓ is a direct consequence of the fact that, in the inertial range, the statistical properties of velocity fluctuations are uniquely and universally determined by the scale ℓ and by the mean energy dissipation rate per unit mass ϵ.

Next we show that dimensional consistency requires that the mean square velocity increment assumes the form

$$\langle [\Delta V(\ell)]^2 \rangle = C \epsilon^{2/3} \ell^{2/3}, \tag{11.7}$$

where C is a dimensionless constant. This equation is the only one possi-

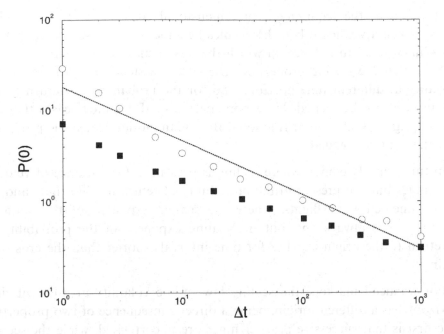

Fig. 11.5. Measured probability of return to the origin of the velocity of a turbulent fluid. Probability of return to the origin $P(0)$ (open circles) and probability of return assuming a Gaussian shape $P_G(0)$ (filled squares) are shown as functions of the time sampling interval Δt. Again, the two measured quantities differ in the full interval, implying that the profile of the PDF must be non-Gaussian. However, in this case, a single scaling power-law behavior does not exist for the entire time interval spanning three orders of magnitude. The slope of the best linear fit (which is of quite poor quality) is -0.59 ± 0.11, while a Gaussian distribution would have slope -0.5. Adapted from [112].

ble because the energy dissipation rate per unit mass has the dimensions $[L]^2[T]^{-3}$. In fact, if we define a to be the exponent of ϵ, and b to be the exponent of ℓ in Eq. (11.7), then dimensional consistency requires that

$$\frac{[L]^2}{[T]^2} = \frac{[L]^{2a}}{[T]^{3a}}[L]^b, \tag{11.8}$$

where the equality indicates that both sides of the equation have the same dimension. This condition is satisfied by equating powers of L and T,

$$\begin{cases} 2 = 2a + b \\ 2 = 3a. \end{cases} \tag{11.9}$$

Hence $a = 2/3$ and $b = 2/3$.

Hence Kolmogorov's law (11.7) is directly related to the observation that the mean energy dissipation rate is the only relevant quantity in the inertial range, and to the requirement of dimensional consistency. In fact, (11.7) loses its validity when other observables become relevant, such as occurs in two-dimensional turbulence, in which there is vorticity conservation.

Note that the scaling properties observed in a stochastic process for the variance at different time horizons and for the probability of return to the origin need not be related. In certain specific (and common) cases they are related, e.g., in Gaussian or fractional Brownian motion stochastic processes. But they are not related

- in turbulent dynamics, where scaling is present in the variance of velocity changes but not present in the probability of return to the origin, and
- in truncated Lévy flights, where the scaling exponent of the variance $\sigma^2(t) \sim t$ is always one, but the scaling exponent of the probability of return to the origin is $-1/\alpha$ for time intervals shorter than the crossover time.

For financial markets, the scaling law of the volatility at different time horizons has a different origin, being a direct consequence of two properties. The first is that successive price changes are uncorrelated, while the second is that the variance of price changes is finite. Hence, unlike turbulence, the scaling property of volatility is related to statistical properties of the underlying stochastic process. Thus we have seen that although scaling can be observed in disparate systems, the *causes* of the scaling need not be the same. Indeed, the fundamental reasons that lead to scaling in turbulence differ from those that lead to scaling in financial markets.

11.4 Discussion

The parallel analysis of velocity fluctuations in turbulence and index (or exchange rate) changes in financial markets shows that the same statistical methods can be used to investigate systems with known, but unsolvable, equations of motion, and systems for which a basic mathematical description of the process is still lacking. In the two phenomena we find both

- *similarities:* intermittency, non-Gaussian pdf, and gradual convergence to a Gaussian attractor in probability, and
- *differences:* the pdfs have different shapes in the two systems, and the probability of return to the origin shows different behavior – for turbulence we do not observe a scaling regime whereas for index changes we observe a scaling regime spanning a time interval of more than three orders of magnitude. Moreover, velocity fluctuations are anticorrelated whereas index (or exchange rate) fluctuations are essentially uncorrelated.

A closer inspection of Kolmogorov's theory explains why the observation of this difference is not surprising. The 2/3 law for the evolution of the variance of velocity fluctuations, Eq. (11.5), is valid only for a system in which the dynamical evolution is essentially controlled by the energy dissipation rate per unit mass. We do not see any rational reason supporting the idea that assets in a financial market should have a dynamical evolution controlled by a similar variable. Indeed no analog of the 2/3 law appears to hold for price dynamics.

12

Correlation and anticorrelation between stocks

One of the more appealing ideas in econophysics is that financial markets can be described along lines similar to successful descriptions of critical phenomena. Critical phenomena are physical phenomena that occur in space (real or abstract) and time. We have considered thus far only a single asset and its time evolution, but in this chapter we discuss an approach based on the simultaneous investigation of several stock-price time series belonging to a given portfolio. Indeed, the presence of cross-correlations (and anticorrelations) between pairs of stocks has long been known, and plays a key role in the theory of selecting the most efficient portfolio of financial goods [49, 115]. We show how relevant these correlations and anticorrelations are by discussing a study devoted to detect the amount of synchronization present in the dynamics of a pair of stocks traded in a financial market [107]. The specific properties of the covariance matrix of stock returns of a given portfolio of stocks have been investigated extensively. Also we briefly consider studies that aim (i) to detect the number of economic factors affecting the dynamics of stock prices in a given financial market [34, 154], and (ii) to evaluate the deviations observed between market data and the results expected from the theory of random matrices [63, 87, 134].

12.1 Simultaneous dynamics of pairs of stocks

In financial markets, many stocks are traded simultaneously. One way to detect similarities and differences in the synchronous time evolution of a pair of stocks is to study the correlation coefficient ρ_{ij} between the daily logarithmic changes in price of two stocks i and j. Generalizing (5.4), we can define for stock i

$$S_i \equiv \ln Y_i(t) - \ln Y_i(t-1), \tag{12.1}$$

98

so

$$\rho_{ij} = \frac{\langle S_i S_j \rangle - \langle S_i \rangle \langle S_j \rangle}{\sqrt{\langle S_i^2 - \langle S_i \rangle^2 \rangle \langle S_j^2 - \langle S_j \rangle^2 \rangle}}. \qquad (12.2)$$

Here Y_i is the daily closure price of stock i at time t, and S_i is the daily change of the logarithm of the price of stock i. The angular brackets indicate a time average over all the trading days within the investigated time period. With this definition, the correlation coefficient ρ_{ij} can assume values ranging from -1 to 1, with three special values

$$\rho_{ij} = \begin{cases} 1 & \text{completely correlated changes in stock price,} \\ 0 & \text{uncorrelated changes in stock price, and} \\ -1 & \text{completely anticorrelated changes in stock price.} \end{cases} \qquad (12.3)$$

We discuss here an investigation of ρ_{ij} carried out for two sets of stocks of the New York Stock Exchange [107].

(i) The 30 stocks used to compute the Dow–Jones Industrial Average (DJIA).

(ii) The 500 stocks used to compute the Standard & Poor's 500 index (S&P 500).

12.1.1 Dow–Jones Industrial Average portfolio

For the set of 30 stocks of the DJIA, there are $(30 \times 29)/2 = 435$ different ρ_{ij}. All the ρ_{ij} are calculated for each investigated time period. Table 12.1 summarizes the minimum and maximum values of the set of ρ_{ij}. From Table 12.1 it is evident that the typical *maximum* value of ρ_{ij} is above 0.5, so in this portfolio there exist some quite positively correlated pairs of stocks. The typical *minimum* value is close to zero, so the degree of maximal anticorrelation is small.

The largest value of the ρ_{ij}, 0.73, is observed in 1990 for the pair of stocks Coca Cola and Procter & Gamble. In Fig. 12.1 the time evolution of $\ln Y(t)$ is shown for both stocks. From the figure it is evident that the prices of the two stocks are remarkably synchronized.

In Table 12.1 only the minimum and maximum values of ρ_{ij} for each time interval are listed. Additional information about the behavior of the correlation coefficient matrix can be obtained by considering the pdf $P(\rho_{ij})$ of the full set of 435 correlation coefficients. In fact, $P(\rho_{ij})$ is a bell-shaped curve; the average is slowly time-dependent, whereas the standard deviation σ is almost constant [107].

Table 12.1. *The observed minimum and maximum*
values when one measures all values of
correlation coefficient ρ_{ij} in the set of 30 stocks
of the Dow–Jones Industrial Average [107].

Time period	Minimum	Maximum
1990	0.02	0.73
1991	−0.01	0.63
1992	−0.10	0.63
1993	−0.16	0.63
1994	−0.06	0.51

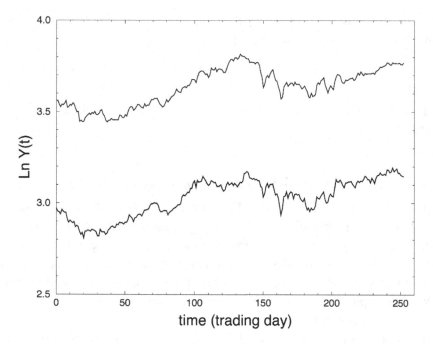

Fig. 12.1. Time evolution of $\ln Y(t)$ for Coca Cola (bottom curve) and Procter & Gamble (top curve) in the year 1990.

For all 435 pairs of stocks, ρ_{ij} changes with time. How long is the characteristic time scale over which strongly correlated pairs of stocks maintain their correlated status? Figure 12.2 shows the time evolution of $\ln Y(t)$ for Coca Cola and Procter & Gamble for the five calendar years investigated.

To quantify the relative value of the correlation coefficient for a pair of

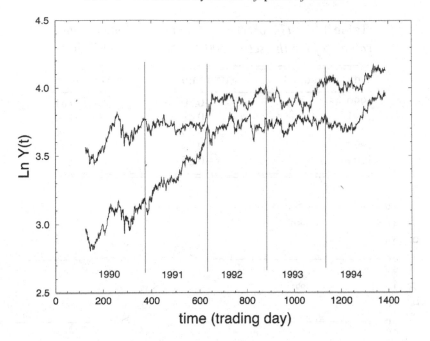

Fig. 12.2. Time evolution of $\ln Y(t)$ for Coca Cola and Procter & Gamble for the five calendar years investigated, 1990 to 1994. The value of ρ_{ij} is 0.73, 0.47, 0.28, 0.33, and 0.39 during the five years from 1990 through 1994, respectively, whereas δ_{ij} is 2.62, 1.73, 1.25, 2.44, and 2.27, respectively, during the same five years.

stocks, we define

$$\delta_{ij} \equiv \frac{\rho_{ij} - \langle \rho_{ij} \rangle}{\sigma} \tag{12.4}$$

to be the deviation of ρ_{ij} from its average value using the standard deviation σ as the unit of measurement, where $\langle \rho_{ij} \rangle$ is the average of ρ_{ij} over all pairs of stocks ij in the portfolio analyzed. For the case where i and j denote Coca Cola and Procter & Gamble, $\delta_{ij} > 1$ for all five years studied, consistent with the possibility that, for this pair of stocks, the correlation coefficient ρ_{ij} varies with a characteristic time scale of years.

12.1.2 S&P 500 portfolio

For the portfolio of stocks used to compute the S&P 500 index, there are $(500 \times 499)/2 = 124{,}750$ different ρ_{ij} – many more than for the 30 stocks included in the DJIA. Table 12.2 lists the minimum and maximum values of ρ_{ij} measured for the S&P 500. Consistent with the results obtained for the DJIA portfolio, we observe pairs of stocks characterized by a high de-

Table 12.2. *The observed minimum and maximum values ρ_{ij} for the set of 500 stocks of the S&P 500.*

Time period	Minimum	Maximum
1990	−0.30	0.81
1991	−0.29	0.74
1992	−0.25	0.73
1993	−0.27	0.81
1994	−0.25	0.82

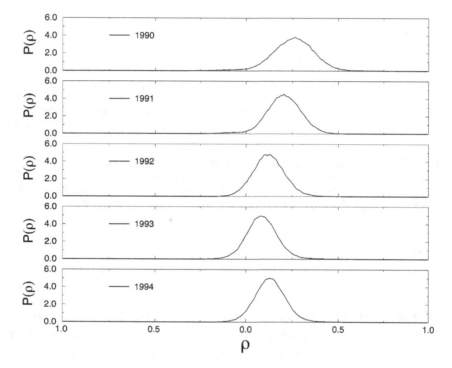

Fig. 12.3. Correlation coefficients for the S&P 500: $P(\rho_{ij})$ is shown for each of the five calendar years 1990 to 1994.

gree of synchronization [107]. The most prominent case is observed in 1994, between Homestake Mining and Placer Dome, Inc. for which $\rho_{ij} = 0.82$. Anticorrelated stocks are also present but, as for the DJIA, the degree of anticorrelation is less than the degree of correlation. The strongest anticorrelation observed – in 1990, between Barrick Gold and Nynex Corporation – is $\rho_{ij} = -0.30$.

Since the total number of correlation coefficients ρ_{ij} is much larger than for the DJIA, the $P(\rho_{ij})$ pdf has a larger statistical reliability. In Fig. 12.3,

$P(\rho_{ij})$ is shown for each of the five calendar years. The figure demonstrates, as for the DJIA case, that the center of the pdf is slowly moving in time, whereas the width is approximately constant.

12.2 Statistical properties of correlation matrices

The statistical properties of the correlation matrix of returns have been investigated in both the economics and the physics literature, but with differing goals. In economics research, a main goal is to determine the number of k factors present in a financial market using the arbitrage pricing theory originally developed by Ross [139]. In this theory, an economic factor is a factor that is common to the set of stocks under consideration; n one-period asset returns R_n are generated by a linear stochastic process with k factors. Specifically,

$$R_n = R_{n0} + B\xi_k + \epsilon_n, \tag{12.5}$$

where R_{n0} represents the risk-free and factor-risk premia mean returns, B the $n \times k$ matrix of factor weights, ξ_k the time series of the k factor affecting the asset returns, and ϵ_n an asset-specific risk. It is assumed that ξ_k and ϵ_n have zero means, and are characterized by covariance $\mathrm{cov}(\xi_k, \xi_\ell) = 0$ when $k \neq \ell$ and $\mathrm{cov}(\xi_i, \epsilon_i) = 0$ for any i.

The statistical properties of the eigenvalues of a random matrix are well documented [39, 69, 119]. Within the framework of the arbitrage pricing theory, the existence of eigenvalues dominating the covariance matrix has been interpreted as evidence of a small number of economic k factors driving the stochastic dynamics of asset returns in a financial market. Empirical analysis seems to suggest that only a few k factors exist, and that there is strong evidence for the existence of a prominent k factor among them [24].

The empirical analyses pursued by physicists also detect a prominent eigenvalue far larger than – and several other eigenvalues slightly larger than – what is expected from random matrix theory [87, 134]. Physicists hope to use the theoretical framework of theories such as Anderson localization theory and spin glass theory to interpret these findings. For example, the Anderson localization theory motivates the monitoring of the lowest eigenvalues, which are associated with eigenvectors that turn out to be controlled by a number of independent elements smaller than for the typical eigenvector [134].

12.3 Discussion

Analyses of the correlation coefficient and of the covariance matrices of asset returns in financial markets show that synchronization between pairs of assets is present in the market. It is plausible that the presence of a relevant degree of cross-correlation between stocks needs to be taken into account in the modeling of financial markets. Evidence of the presence of a small number of economic factors driving a large number of assets is also detected. These findings are not inconsistent with the efficient market hypothesis because synchronization between assets and the existence of economic factors do not directly imply the temporal predictability of future asset prices. Cross-correlations after a given time lag, and a precise knowledge of the nature of factors and their dynamics, if present, would provide arbitrage opportunities and deviation from market efficiency. Indeed, one of these deviations has been detected by observing that returns of large stocks lead those of smaller stocks [97].

13

Taxonomy of a stock portfolio

In Chapter 12, we introduced the notion of a correlation coefficient ρ_{ij} to quantify the degree of synchronization of stock i and stock j. In this chapter, we will see that this concept is useful in two different ways: (i) it allows us to define a metric that provides the relative distance between the stocks of a given portfolio, and (ii) it provides a method for extracting economic information stored in the stock-price time series.

13.1 Distance between stocks

A method of determining a distance between stocks i and j evolving in time in a synchronous fashion is the following. Let us consider

$$\tilde{S}_i \equiv \frac{S_i - \langle S_i \rangle}{\sqrt{\langle S_i^2 \rangle - \langle S_i \rangle^2}}, \tag{13.1}$$

where S_i, the logarithmic price difference of stock i, is given by Eq. (12.1). Hence $\tilde{S}_i$ is the same variable subtracted from its mean, and divided by its standard deviation computed over a given time interval. Let us consider the n records of $\tilde{S}_i$ present in the same time interval as the components $\tilde{S}_{ik}$ of an n-dimensional vector $\tilde{\mathbf{S}}_i$. The Euclidean distance d_{ij} between vectors $\tilde{\mathbf{S}}_i$ and $\tilde{\mathbf{S}}_j$ is obtainable from the Pythagorean relation

$$d_{ij}^2 = ||\tilde{\mathbf{S}}_i - \tilde{\mathbf{S}}_j||^2 = \sum_{k=1}^{n} (\tilde{S}_{ik} - \tilde{S}_{jk})^2. \tag{13.2}$$

The vector $\tilde{\mathbf{S}}_i$ has unit length because, from definition (13.1),

$$\sum_{k=1}^{n} \tilde{S}_{ik}^2 = 1. \tag{13.3}$$

Hence Eq. (13.2) can be rewritten as

$$d_{ij}^2 = \sum_{k=1}^{n}(\tilde{S}_{ik}^2 + \tilde{S}_{jk}^2 - 2\tilde{S}_{ik}\tilde{S}_{jk}) = 2 - 2\sum_{k=1}^{n}\tilde{S}_{ik}\tilde{S}_{jk}. \qquad (13.4)$$

The sum on the right side of Eq. (13.4), $\sum_{k=1}^{n}\tilde{S}_{ik}\tilde{S}_{jk}$, coincides with ρ_{ij} (see Eq. (12.2)). Hence Eq. (13.4) leads to (D. Sornette, private communication)

$$d_{ij} = \sqrt{2(1 - \rho_{ij})}. \qquad (13.5)$$

Because Eq. (13.2) defines a Euclidean distance, the following three properties must hold:

$$
\begin{aligned}
\text{Property (i)} : \quad & d_{ij} = 0 \iff i = j \\
\text{Property (ii)} : \quad & d_{ij} = d_{ji} \qquad\qquad\qquad (13.6) \\
\text{Property (iii)} : \quad & d_{ij} \le d_{ik} + d_{kj}
\end{aligned}
$$

Properties (i) and (ii) are easily verified because $\rho_{ij} = 1$ implies $d_{ij} = 0$, while $\rho_{ij} = \rho_{ji}$ implies $d_{ij} = d_{ji}$. The validity of Property (iii), the 'triangular inequality', relies on the equivalence of Eq. (13.2) and Eq. (13.5). Thus the quantity d_{ij} fulfills all three properties that must be satisfied by a metric distance.

The introduction of a distance between a synchronous evolving pair of assets was first proposed in [108], where a distance numerically verifying properties (i)–(iii) was used. The knowledge of the distance matrix between n objects is customarily used to decompose the set of n objects into subsets of closely related objects. To obtain such a taxonomy, an additional hypothesis about the topological space of n objects needs to be formed, and this is the subject of the next section.

13.2 Ultrametric spaces

Consider a specific example, a portfolio of $n = 6$ stocks: Chevron (CHV), General Electric (GE), Coca Cola (KO), Procter & Gamble (PG), Texaco (TX), and Exxon (XON), where in parentheses we identify their tick symbols. Starting from the measured values of ρ_{ij} over the calendar year 1990, we calculate the distance matrix d_{ij}

	CHV	GE	KO	PG	TX	XON
CHV	0	1.15	1.18	1.15	0.84	0.89
GE		0	0.86	0.89	1.26	1.16
KO			0	0.74	1.27	1.11
PG				0	1.26	1.10
TX					0	0.94
XON						0

We make the working hypothesis that a useful space for linking n stocks is an ultrametric space. This hypothesis is motivated *a posteriori* by the fact that the associated taxonomy is meaningful from an economic point of view. An ultrametric space is a space in which the distance between objects is an ultrametric distance. An ultrametric distance $\hat{d}_{ij}$ must satisfy the first two properties of a metric distance, (i) $\hat{d}_{ij} = 0 \iff i = j$ and (ii) $\hat{d}_{ij} = \hat{d}_{ji}$, while the usual triangular inequality of Eq. (13.6) is replaced by a stronger inequality, called an ultrametric inequality,

$$\hat{d}_{ij} \leq \max\{\hat{d}_{ik}, \hat{d}_{kj}\}. \tag{13.7}$$

Ultrametric spaces provide a natural way to describe hierarchically structured complex systems, since the concept of ultrametricity is directly connected to the concept of hierarchy. They are observed in spin glasses [123], the archetype of frustrated disordered systems. A good introduction to the concept of ultrametricity for the reader with a background in physical science is provided by Rammal *et al.* [138].

The general connection between indexed hierarchies and ultrametrics was rigorously studied by Benzécri [15]. Provided that a metric distance between n objects exists, several ultrametric spaces can be obtained by performing any given partition of the set of n objects. Among all the possible ultrametric structures associated with the distance metric, d_{ij}, a single one emerges owing to its simplicity and remarkable properties. This is the subdominant ultrametric. In the presence of a metric space in which n objects are linked together, the subdominant ultrametric can be obtained by determining the minimal-spanning tree (MST) connecting the n objects. The MST is a concept in graph theory [157]. In a connected weighted graph of n objects, the MST is a tree having $n - 1$ edges that minimize the sum of the edge distances. The subdominant ultrametric space associated with a metric space provides a well defined topological arrangement that has associated a unique indexed hierarchy. Hence the investigation of the subdominant ultrametrics allows one to determine in a unique way an indexed hierarchy of the n objects

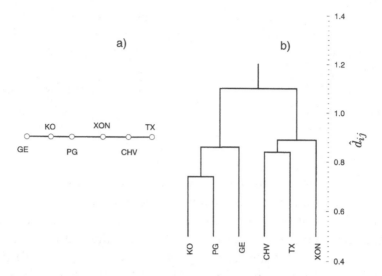

Fig. 13.1. (a) MST and (b) indexed hierarchical tree obtained for the example of six firms, identified by their tick symbols CHV, GE, KO, PG, TX and XON.

considered. The method of constructing a MST linking a set of n objects, known as Kruskal's algorithm [130, 157], is simple and direct.

The MST associated with the Euclidean matrix can be obtained as follows. First find the pair of stocks separated by the smallest distance: KO and PG ($d = 0.74$). Then find the pair of stocks with the next-smallest distance: CHV and TX ($d = 0.84$). We now have two separate regions in the MST. If we continue, we find next the KO and GE pair ($d = 0.86$). At this point, the regions of the MST are GE–KO–PG and CHV–TX. The next pairs of closest stocks are GE–PG and CHV–XON ($d = 0.89$). The connection GE–PG is not considered because both stocks have already been sorted, while XON is linked to CHV in the MST. Now the two regions are XON–CHV–TX and GE–KO–PG. The smallest distance connecting the two regions is observed for PG–XON ($d = 1.10$). This PG–XON link completes the MST.

Using this procedure, it is possible to obtain the MST shown in Fig. 13.1a. In Fig. 13.1b we show the indexed hierarchical tree associated with the MST. The tree shows clearly that in this portfolio there are two groups of stocks. In the first group are the oil companies (CHV, TX, and XON), and in the second are companies selling consumer products or consumer services (KO, PG, and GE). If we start from the indexed hierarchical tree, determining the matrix of the ultrametric distance $\hat{d}_{ij}$ is straightforward. In our example, the $\hat{d}_{ij}$ matrix is

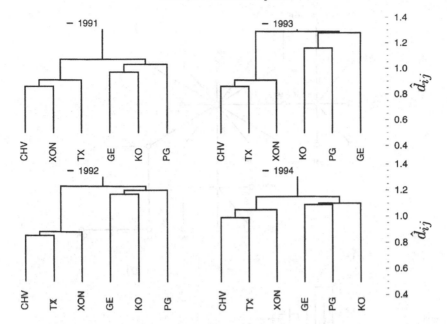

Fig. 13.2. Indexed hierarchical trees obtained during the calendar years from 1991 to 1994 for the portfolio of six firms (CHV, GE, KO, PG, TX, and XON).

	CHV	GE	KO	PG	TX	XON
CHV	0	1.10	1.10	1.10	0.84	0.89
GE		0	0.86	0.86	1.10	1.10
KO			0	0.74	1.10	1.10
PG				0	1.10	1.10
TX					0	0.89
XON						0

Each element in the $\hat{d}_{ij}$ matrix is equal to the maximal distance between two successive objects encountered when moving from the starting object to the ending object over the shortest path of the MST connecting the two objects. In contrast to the d_{ij} matrix, the number of different element values in the ultrametric distance matrix $\hat{d}_{ij}$ cannot exceed $n-1$, as is confirmed by the present example.

In Chapter 12, we showed that the time evolution of ρ_{ij} can be characterized by slow dynamics over a time scale of years. However, ρ_{ij} is a statistical quantity and it is relevant to consider how stable a hierarchical structure can be (Fig. 13.1b). In Fig. 13.2 we show the indexed hierarchical trees obtained in the calendar years from 1991 to 1994 for the portfolio of six stocks discussed above.

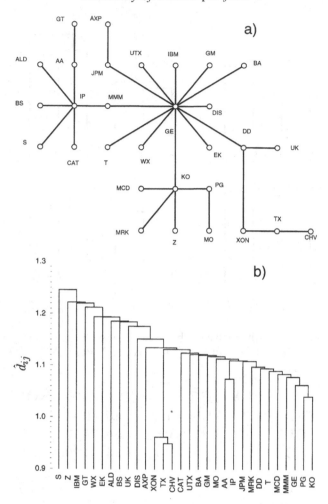

Fig. 13.3. (a) MST and (b) indexed hierarchical tree obtained for the DJIA portfolio during the time period 7/89 to 10/95. Adapted from [108].

The two main clusters observed in 1990 (Fig. 13.1b), CHV–TX–XON and GE–PG–KO, are also observed in all the other years. But the value of the baseline distance is time-dependent and the internal structure of the two clusters varies. For example, in four of the five years the closest oil companies are CHV and TX, whereas in 1991 the closest are CHV and XON (cf. Figs. 13.1 and 13.2). The most strongly connected consumer product companies are KO and PG in 1990 and 1993, GE and KO in 1991, and GE and PG in 1994.

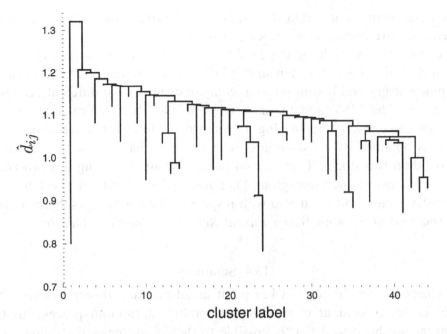

Fig. 13.4. Main structure of the MST of the S&P 500 portfolio for the time period 7/89 to 10/95. Adapted from [108].

In summary, empirical analyses show that the indexed hierarchical tree is time-dependent, but maintains on a time scale of years a basic structure that exhibits a meaningful economic taxonomy.

13.3 Subdominant ultrametric space of a portfolio of stocks

The procedure outlined above has been used [108] to obtain the ultrametric space of two stock portfolios. The first is the set of 30 stocks used to compute the Dow–Jones Industrial Average (DJIA) index. Figure 13.3 shows the MST obtained for the DJIA portfolio during the time period 7/89 to 10/95, as well as the associated indexed hierarchical tree. Three groups of linked stocks are seen in this figure. The first group is made up of oil companies (CHV, TX, and XON), the second of consumer-product or consumer-service companies (PG, KO, GE, MMM, MCD, T, DD, MRK, and JPM), and the third of raw-material companies (IP and AA). The taxonomy associated with the subdominant ultrametric of the DJIA portfolio is a meaningful economic taxonomy. Notice that this taxonomy is obtained by starting from the time series of stock prices, without any assumptions other than the fundamental one – that the subdominant ultrametric well describes the reciprocal arrangement of a stock portfolio.

The second portfolio is the S&P 500. The associated taxonomy is more refined in this portfolio than in the DJIA, because the S&P 500 portfolio is much larger and because several companies in it have different economic activities. The MST, and the associated indexed hierarchical tree, are too complex to display here, but Fig. 13.4 shows the main structure of the MST. This is obtained by considering only those lines that end in a group of no fewer than two stocks. There are 44 groups obtainable using this procedure and, in most cases, these groups are homogeneous with respect to their industry sector, and often also with respect to their industry subsector [108] as specified in the 49th *Forbes Annual Report on American Industry*.

13.4 Summary

In Chapter 2 we discussed a key point in information theory: a time series that is not redundant often closely resembles a random process. In this chapter, we have seen that is possible to devise strategies that allow us to obtain meaningful taxonomies if we start from the synchronous analysis of more than one stock-price time series. Specifically, we can retrieve part of the economic information stored in the individual stock-price time series if we calculate the distance between each pair of stocks in a portfolio, and we assume that a subdominant ultrametric space is an appropriate topology.

14

Options in idealized markets

In the previous chapters we have seen that the dynamics of stock prices is a complex subject, and that a definitive model has yet to be constructed. The complexity of the entire financial system is even greater. Not only is the trading of financial securities complex, but additional sources of complexity come from the issuing of financial contracts on those fluctuating financial securities.

An important class of financial contracts is derivatives, a financial product whose price depends upon the price of another (often more basic) financial product [22, 45, 73, 74, 122, 127]. Examples of derivatives include forward contracts, futures, options, and swaps. Derivatives are traded either in over-the-counter markets or, in a more formalized way, in specialized exchanges. In this chapter, we examine the most basic financial contracts and procedures for their rational pricing. We consider idealized markets and we discuss the underlying hypothesis used in obtaining a rational price for such a contract.

14.1 Forward contracts

The simplest derivative is a *forward contract*. When a forward contract is stipulated, one of the parties agrees to buy a given amount of an asset at a specified price (called the forward price or the delivery price K) on a specified future date (the delivery date T). The other party agrees to sell the specified amount of the asset at the delivery price on the delivery date. The party agreeing to buy is said to have a *long position*, and the party agreeing to sell is said to have a *short position*.

The actual price Y of the underlying financial asset fluctuates, and the price $Y(T)$ at the delivery date usually differs from the delivery price specified in the forward contract. The payoff is either positive or negative, so whatever is gained by one party will be lost by the other (Fig. 14.1).

113

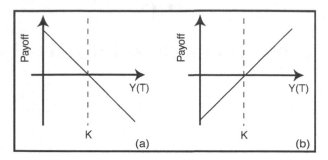

Fig. 14.1. Payoff for each party involved in a forward contract, as a function of the price $Y(T)$ at maturity time T, (a) for the short position (party which agrees to sell in the future), (b) for the long position (party which agrees to buy in the future).

14.2 Futures

A *future contract* is a forward contract traded on an exchange. This implies that the contract is standardized and that the two parties interact through an exchange institution, the *clearing house*. Immediately following the completion of the trade, the clearing house writes the contracts – one with the buyer and one with the seller. The clearing house guarantees that its contracts will be executed at the delivery date.

14.3 Options

An option is a financial contract that gives the holder the right to exercise a given action (buying or selling) on an underlying asset at time T and at price K. The price K is called the strike price or the exercise price, and T is called the expiration date, the exercise date, or the date of maturity.

Options can also be characterized by the nature of the period during which the option can be exercised. If the option can be exercised only at maturity, $t = T$, it is called a European option. If the option can be exercised at any time between the contract initiation at $t = 0$ and $t = T$, it is called an American option. In this chapter we consider European options.

There are call options and put options. In a *call* option, the buyer of the option has the right to buy the underlying financial asset at a given strike price K at maturity. This right is obtained by paying to the seller of the option an amount of money $C(Y, t)$. In a call option there is no symmetry between the two parties of the contract. The buyer of the option pays money when the contract is issued and acquires thereby a right to be exercised in the future, while the seller of the option receives cash immediately but faces potential liabilities in the future (Fig. 14.2). In a *put* option, the

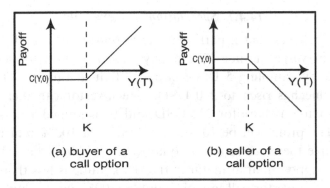

Fig. 14.2. (a) The payoff as a function of the price at maturity time T for a buyer of a call option, which costs $C(Y,t)$ at the time the contract is written, where K denotes the strike price. (b) The payoff for the seller of the same call option.

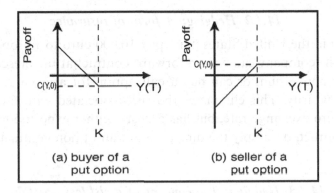

Fig. 14.3. Same as Fig. 14.2, for a put option.

buyer of the option has the right to sell the underlying financial asset at a given strike price K at maturity $(t = T)$ back to the seller of the option (Fig. 14.3).

14.4 Speculating and hedging

Derivatives are attractive financial products for at least two types of traders: speculators and hedgers. *Speculators* are interested in derivatives because they can provide an inexpensive way to expose a portfolio to a large amount of risk. *Hedgers* are interested in derivatives because they allow investors to reduce the market risk to which they are already exposed.

14.4.1 Speculation: An example

An investor believes that a particular stock (currently priced at 200 USD) will go up in value at time T. He buys a call option for a strike price of 220 USD by gambling 5 USD. Suppose that at time $t = T$ (maturity), the stock value has risen to 230 USD. The investor can then exercise his option by buying a share for 220 USD and then immediately selling it for 230 USD. The profit will be $10 - 5 = 5$ USD – a 100% return. Note that in this example the stock return is equal to $(230 - 200)/200 = 15\%$. On the other hand, suppose that at maturity the stock value is less than or equal to 220 USD – the investor will lose his gamble (100% loss). Thus the investor, by gambling 5 USD, becomes eligible for huge returns at the expense of exposing himself to huge risks.

14.4.2 Hedging: A form of insurance

A company in the United States must pay 10,000 euro to European firms in 180 days. The company can write a forward contract at the present exchange rate for the above sum, or can buy a call option for a given strike price at 180 days' maturity. This eliminates the risk associated with fluctuations in the USD/euro exchange rate, but has a cost – either exposure to losses in a forward contract, or simply the direct cost in an option contract.

14.4.3 Hedging: The concept of a riskless portfolio

To examine more closely the procedure of hedging, we consider a simplified version of our problem, namely a binomial model of stock prices [37]. The price Y at each time step t may assume only two values (Fig. 14.4). Suppose a hedger at each time step holds a number Δ_h of shares for each option sold on the same stock. In order to minimize the risk, the hedger needs to determine the value of Δ_h that makes the portfolio riskless. The value ϕ of a portfolio is

$$\phi = Y\Delta_h - C, \tag{14.1}$$

where $Y\Delta_h$ is the value of Δ_h shares held by the investor at time t, and C is the value of the option sold at time t.

A riskless investment requires

$$\phi_u = \phi_d, \tag{14.2}$$

where ϕ_u is the portfolio value if the stock goes up, while ϕ_d is the portfolio

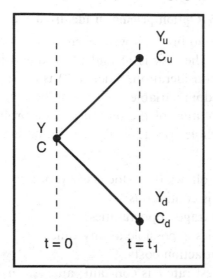

Fig. 14.4. Schematic illustration of the binomial model. Here Y denotes the stock share price, where C denotes the cost of an option issued on the underlying stock. For a given time horizon $t = t_1$, Y_u and Y_d denote the possible stock price values, while C_u and C_d denote the possible values of options.

value if the stock goes down. Hence from (14.1),

$$Y_u \Delta_h - C_u = Y_d \Delta_h - C_d,$$

or

$$\Delta_h = \frac{C_u - C_d}{Y_u - Y_d}. \tag{14.3}$$

In the limit when t becomes infinitesimal

$$\Delta_h = \frac{\partial C}{\partial Y}. \tag{14.4}$$

Thus Δ_h is equal to the partial derivative of the price of the option with respect to the price of the stock (at constant t). Because Y changes over time, $\Delta_h = \partial C/\partial Y$ must also be changed over time in order to maximize the effectiveness of the hedging and minimize the risk to the portfolio.

We have seen that at least three different trading strategies are used in financial markets: hedging, speculating, and exploiting arbitrage opportunities. Some traders specialize in one of these three, while others occasionally switch from strategy to strategy. Hedgers focus on portfolio risk reduction, while speculators maximize portfolio risk.

14.5 Option pricing in idealized markets

For a financial market to function well, participants must thoroughly understand option pricing. The task is to find the rational and fair price $C(Y,t)$ of the option under consideration. Since $Y(t)$ is a random variable, $C(Y,t)$ is a function of a random variable.

The first reliable solution of the option-pricing problem was proposed in 1973 by Black and Scholes [18, 120]. Their solution is valid under a series of assumptions:

(i) The stock price follows Ito's stochastic process;
(ii) security trading is continuous;
(iii) there are no arbitrage opportunities;
(iv) selling of securities is possible at any time;
(v) there are no transaction costs;
(vi) the market interest rate r is constant; and
(vii) there are no dividends between $t = 0$ and $t = T$.

Black and Scholes assume that a stock price $Y(t)$ can be described as an Ito process, namely a process defined by the stochastic differential equation $dY = a(Y,t)dt + b(Y,t)dW$. Specifically, they assume that a stock price follows a geometric Brownian motion

$$dY = \mu Y\, dt + \sigma Y\, dW, \tag{14.5}$$

where μ is the expected return per unit time, σ^2 the variance per unit time, and W a Wiener process. This assumption implies that the changes in the logarithm of price are Gaussian distributed.

If one assumes that a stock price is modeled by a geometric Brownian motion, any function of Y (including the price of the option C) must be a solution of the partial differential equation obtained from a special case of Ito's lemma valid for a geometric Brownian motion [75],

$$dC = \left[\frac{\partial C}{\partial Y}\mu Y + \frac{\partial C}{\partial t} + \frac{1}{2}\frac{\partial^2 C}{\partial Y^2}\sigma^2 Y^2\right] dt + \frac{\partial C}{\partial Y}\sigma Y\, dW. \tag{14.6}$$

Let us consider the portfolio of the holder of Δ_h shares who is selling one derivative of the stock at time t. The value of the portfolio is, from (14.1) and (14.4),

$$\phi = -C + \frac{\partial C}{\partial Y}Y. \tag{14.7}$$

The change in the value of the portfolio ϕ over a time interval Δt is

$$\Delta\phi = -\Delta C + \frac{\partial C}{\partial Y}\Delta Y. \tag{14.8}$$

Using Ito's lemma, we have

$$\Delta C = \left(\frac{\partial C}{\partial Y} \mu Y + \frac{\partial C}{\partial t} + \frac{1}{2} \frac{\partial^2 C}{\partial Y^2} \sigma^2 Y^2 \right) \Delta t + \frac{\partial C}{\partial Y} \sigma Y \Delta W. \tag{14.9}$$

From the definition of geometric Brownian motion, (14.5), we have

$$\Delta Y = \mu Y \Delta t + \sigma Y \Delta W. \tag{14.10}$$

Hence the change in ϕ is

$$\Delta \phi = \left[-\frac{\partial C}{\partial Y} \mu Y - \frac{\partial C}{\partial t} - \frac{1}{2} \frac{\partial^2 C}{\partial Y^2} \sigma^2 Y^2 + \frac{\partial C}{\partial Y} \mu Y \right] \Delta t$$

$$- \frac{\partial C}{\partial Y} \sigma Y \Delta W + \frac{\partial C}{\partial Y} \sigma Y \Delta W, \tag{14.11}$$

which simplifies to

$$\Delta \phi = \left[-\frac{\partial C}{\partial t} - \frac{1}{2} \frac{\partial^2 C}{\partial Y^2} \sigma^2 Y^2 \right] \Delta t. \tag{14.12}$$

The Black & Scholes assumption that a stock price follows a geometric Brownian motion turns out to be crucial in deriving the rational price of an option. In fact, without this assumption, $\Delta \phi$ could not be simplified as it is in Eq. (14.12).

The second key assumption concerns the absence of arbitrage. In the absence of arbitrage opportunities, the change in the value of portfolio $\Delta \phi$ must equal the gain obtained by investing the same amount of money in a riskless security that provides a return per unit of time r. Under the assumption that r is constant,

$$\Delta \phi = r \phi \Delta t. \tag{14.13}$$

By equating the two equations for the change in the portfolio value, (14.12) and (14.13), we obtain

$$rC = \frac{\partial C}{\partial t} + rY \frac{\partial C}{\partial Y} + \frac{1}{2} \frac{\partial^2 C}{\partial Y^2} \sigma^2 Y^2, \tag{14.14}$$

which is called the Black & Scholes partial differential equation. To obtain (14.14), no assumption about the specific kind of option has been made. This partial differential equation is valid for both call and put European options.

The appropriate $C(Y, t)$ for the chosen type of option will be obtained by selecting the appropriate boundary conditions. An example for a call option is

$$C = \max\{Y - K, 0\} \quad \text{when} \quad t = T. \tag{14.15}$$

The parameters in Eq. (14.14) are the variance per unit of time σ^2 and the return per unit of time r of the riskless security. The solution of (14.14) depends on these two parameters, and on the values of Y, K, and T characterizing the boundary conditions. The Black & Scholes partial differential equation has an analytic solution, which is discussed in the next section.

14.6 The Black & Scholes formula

Black and Scholes solved their partial differential equation (14.14) by making the following substitution

$$C(Y, t) = e^{r(t-T)} y(x, t'), \tag{14.16}$$

where

$$x \equiv \frac{2}{\sigma^2} \left(r - \frac{1}{2}\sigma^2 \right) \left[\ln\left(\frac{Y}{K}\right) - \left(r - \frac{\sigma^2}{2} \right)(t - T) \right], \tag{14.17}$$

and

$$t' \equiv -\frac{2}{\sigma^2} \left(r - \frac{\sigma^2}{2} \right)(t - T). \tag{14.18}$$

With this substitution, the Black & Scholes partial differential equation becomes formally equivalent to the heat-transfer equation of physics,

$$\frac{\partial y(x, t')}{\partial t'} = \frac{\partial^2 y(x, t')}{\partial x^2}. \tag{14.19}$$

The heat-transfer equation is analytically solvable and, by using substitutions (14.17) and (14.18), Black and Scholes found their famous equation for the option-pricing problem,

$$C(Y, t) = Y N(d_1) - K e^{r(t-T)} N(d_2), \tag{14.20}$$

where $N(x)$ is the cumulative density function for a Gaussian variable with zero mean and unit standard deviation,

$$d_1 \equiv \frac{\ln(Y/K) + (r + \sigma^2/2)(T - t)}{\sigma\sqrt{T - t}}, \tag{14.21}$$

and

$$d_2 = d_1 - \sigma\sqrt{T - t}. \tag{14.22}$$

14.7 The complex structure of financial markets

The Black & Scholes model provides two important financial instruments: (a) an analytic solution (14.20) for the rational price of a European option, and (b) a trading strategy for building up a riskless portfolio. The existence of a riskless portfolio implies also that a specific portfolio of bonds and underlying stock can be equivalent to an option issued on the underlying stock at any time if the portfolio is properly balanced in terms of Eqs. (14.1) and (14.4). In other words, the value of an option can be replicated by an appropriate portfolio of stocks and bonds, and synthetic options can be realized in a financial market obeying the Black & Scholes assumptions.

Among assumptions (i) and (vii) of the Black & Scholes model, two assumptions are crucial to the existence of a riskless portfolio. The first is that the path of stock price dynamics is a geometric Brownian motion. The second is that security trading is continuous in time.

In the previous chapters, we saw that the ultimate dynamics of a stock price are discrete in both time and space. Moreover, empirical observations of the statistical properties of price-change statistics for the high-frequency regime do not support the geometric Brownian motion assumption. Indeed, rare events (namely large jumps in the price of a given stock) are observed from time to time.

Hence the Black & Scholes model is a beautiful framework for understanding and modeling an ideal financial market, but provides only an approximate description of real financial markets. In particular, the Black & Scholes assumptions are not verified in real markets and they do not guarantee the existence of a riskless portfolio and of synthetic options in real markets.

14.8 Another option-pricing approach

Other aspects of the option-pricing problem emerge by considering an alternative way of obtaining the rational price of an option. In a Black & Scholes market, the rational price of an option does not depend on the risk tolerance of economic agents. This implies that the assumption of risk-neutrality is legitimate in a financial market without imperfections. In a risk-neutral economy, the expected rate of return μ of an underlying financial asset must be equal to the interest rate r. Hence, in the absence of arbitrage opportunities, the expected value of an European call option at maturity ($t = T$) is the average expected payoff, namely $E\{(Y(T) - K)^+\}$, where $(Y(T) - K)^+$ is $Y(T) - K$ when $Y(T) - K > 0$ and zero when $Y(T) - K \leq 0$. To obtain the risk-neutral value valid at time $T - t$, this value needs to be discounted at the risk-free interest rate, so

$$C(Y,t) = e^{-r(T-t)} E\{(Y(T) - K)^+\}. \tag{14.23}$$

Equation (14.23) provides a tool for determining $C(Y,t)$ without solving the partial differential equation associated with the option-pricing problem. Equation (14.23) can be written in explicit form as

$$C(Y,t) = e^{-r(T-t)} \int_K^\infty dY'(Y' - K) f(Y_0;0 \mid Y';T), \tag{14.24}$$

where $f(Y_0;0 \mid Y';T)$ is the conditional probability density of observing $Y = Y'$ at time $t = T$ when $Y = Y_0$ at time $t = 0$

Equation (14.24) shows how crucial exact knowledge of the price-change distribution is. Indeed, $C(Y,t)$ is controlled completely by its exact shape in a financial market without imperfections.

Equation (14.24) also provides a flexible tool for the analytic or numerical determination of $C(Y,t)$ when the distribution of price changes is known. However, it is worth pointing out that Eq. (14.24) is valid only when the stochastic process of $Y(t)$ allows the building of a riskless portfolio under a risk-neutrality assumption. If there is no riskless portfolio under a risk-neutrality assumption, there is also no guarantee that a unique option price exists satisfying the condition that arbitrage opportunities are not present.

14.9 Discussion

The Black & Scholes solution of the option-pricing problem is a milestone in modern finance. Their model of financial activity catches the basic features of real financial markets. Some aspects, however, do not fully reflect the stochastic behavior observed in real markets. To cite three examples: (i) the Gaussian hypothesis of changes in the logarithm of a stock price is incorrect – especially when changes are high frequency; (ii) the path of the underlying asset price can be discontinuous at the arrival of relevant economic information; and (iii) the volatility of a given stock or index and the interest rate are not constant, and are themselves random processes. The modeling of real financial markets, sometimes called the modeling of 'markets with imperfections', involves a class of problems that we introduce in the next chapter.

15
Options in real markets

In Chapter 14, we considered the option-pricing problem in ideal friction-less markets. Real markets are often efficient, but they are never ideal. In this chapter, we discuss how the complexity of modeling financial markets increases when we take into account aspects of real markets that are not formalized in the ideal model. These aspects are addressed in the literature as market microstructure [26] or market imperfections [127].

The terminology used in the economics literature suggests a clear parallel with similar scenarios observed in physical sciences. For example, it is much easier to construct a generalized description of the motion of a mechanical system in an idealized world without friction than in the real world. A similar situation is encountered when we compare equilibrium and non-equilibrium thermodynamics. In this chapter, we show that knowledge of the statistical properties of asset price dynamics is crucial for modeling real financial markets. We also address some of the theoretical and practical problems that arise when we take market imperfections into account.

15.1 Discontinuous stock returns

The existence of a portfolio containing both riskless and risky assets – replicating exactly the value of an option – is essential in determining the rational price of the option under the assumption that no arbitrage opportunities are present. Whether a portfolio is replicating or not depends on the statistical properties of the dynamics of the underlying asset. In the previous chapter, we saw that a replicating portfolio exists when the price of the underlying asset follows a geometric Brownian motion, but we also saw that this case cannot be generalized. For example, when the asset dynamic follows a jump-diffusion model [121], a simple replicating portfolio does not exist. A jump-diffusion model is a stochastic process composed of a diffusive term (as in geometric Brownian motion) plus a second term describing jumps of random amplitudes occurring at random times.

Roughly speaking, the presence of two independent sources of randomness in the asset price dynamics does not allow the building of a simple replicating portfolio.† It is not possible to obtain the rational price of an option just by assuming the absence of arbitrage opportunities. Other assumptions must be made concerning the risk aversion and price expectations of the traders.

Taking a different perspective, we can say that we need to know the statistical properties of a given asset's dynamics before we can determine the rational price of an option issued on that asset. Discontinuity in the path of the asset's price is only one of the 'imperfections' that can force us to look for less general option-pricing procedures.

15.2 Volatility in real markets

Another 'imperfection' of real markets concerns the random character of the volatility of an asset price. The Black & Scholes option-pricing formula for an European option traded in an ideal market depends only on five parameters: (i) the stock price Y at time t, (ii) the strike price K, (iii) the interest rate r, (iv) the asset volatility rate σ, and (v) the maturity time T. Of these parameters, K and T are set by the kind of financial contract issued, while Y and r are known from the market. Thus the only parameter that needs to be determined is the volatility rate σ.

Note that the volatility rate needed in the Black & Scholes pricing formula is the volatility rate of the underlying security that will be observed in the future time interval spanning $t = 0$ and $t = T$. A similar statement can be made about the interest rate r, which may jump at future times.

We know from the previous analysis that the volatility of security prices is a random process. Estimating volatility is not a straightforward procedure.

15.2.1 Historical volatility

The first approach is to determine the volatility from historical market data. Empirical tests show that such an estimate is affected by the time interval used for the determination. One can argue that longer time intervals should provide better estimations. However, the local nonstationarity of the volatility versus time implies that unconditional volatility, estimated by using very long time periods, may be quite different from the volatility observed in the lifetime of the option.

† For a more rigorous discussion of this point, see [44, 70].

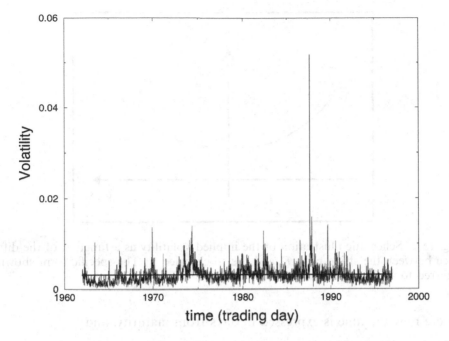

Fig. 15.1. Schematic illustration of the problems encountered in the determination of historical volatility. The nonstationary behavior of the volatility makes the determination of the average volatility depend on the investigated period of time. Long periods of time are observed when the daily volatility is quite different from the mean asymptotic value (solid line).

An empirical rule states that the best estimate of volatility rate is obtained by considering historical data in a time interval $t_1 - t_2$ chosen to be as long as the time to maturity T of the option (Fig. 15.1).

15.2.2 Implied volatility

A second, alternative approach to the determination of the volatility is to estimate the implied volatility σ_{imp}, which is determined starting from the options quoted in the market and using the Black & Scholes option-pricing formula (14.20). The implied volatility gives an indication about the level of volatility expected for the future by options traders.

The value of σ_{imp} is obtained by using the market values of $C(Y, t)$ and by solving numerically the equation

$$C(Y, T - t) = Y N(d_1) - K e^{-r(T-t)} N(d_2),$$ (15.1)

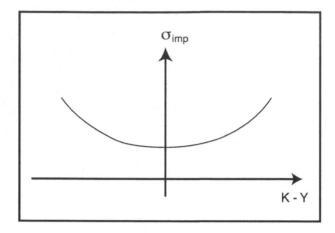

Fig. 15.2. Schematic illustration of the implied volatility as a function of the difference between the strike price K and the stock price Y. The specific form shown is referred to as a volatility smile.

where now the time is expressed in days from maturity, and

$$d_1 \equiv \frac{\ln(Y/K) + (r + \sigma_{\text{imp}}^2/2)(T-t)}{\sigma_{\text{imp}}\sqrt{T-t}}, \tag{15.2}$$

and

$$d_2 \equiv d_1 - \sigma_{\text{imp}}\sqrt{T-t}. \tag{15.3}$$

In a Black & Scholes market, a determination of the implied volatility rate would give a constant value σ for options with different strike prices and different maturity. Moreover, the value of the implied volatility should coincide with the volatility obtained from historical data.

In real markets, the two estimates, in general, do not coincide. Implied volatility provides a better estimate of σ. Empirical analysis shows that σ_{imp} is a function of the strike price and of the expiration date. Specifically, σ_{imp} is minimal when the strike price K is equal to the initial value of the stock price Y ('at the money'), and increases for lower and higher strike prices. This phenomenon is often termed a 'volatility smile' (Fig. 15.2). The implied volatility increases when the maturity increases. These empirical findings confirm that the Black & Scholes model relies on assumptions that are only partially verified in real financial markets.

When random volatility is present, it is generally not possible to determine the option price by simply assuming there are no arbitrage opportunities. In some models, for example, the market price of the volatility risk needs to be specified before the partial differential equation of the option price can be obtained.

15.3 Hedging in real markets

In idealized financial markets, the strategy for perfectly hedging a portfolio consisting of both riskless and risky assets is known. In real markets, some facts make this strategy unrealistic: (i) the rebalancing of the hedged portfolio is not performed continuously; (ii) there are transaction costs in real markets; (iii) financial assets are often traded in round lots of 100 and assume a degree of indivisibility.

It has been shown that the presence of these unavoidable market imperfections implies that a perfect hedging of a portfolio is not guaranteed in a real market, even if one assumes that the asset dynamics are well described by a geometric Brownian motion [58]. When we consider real markets, the complexity of the modeling grows, the number of assumptions increases, and the generality of the solutions diminishes.

15.4 Extension of the Black & Scholes model

It is a common approach in science to use a model system to understand the basic aspect of a scientific problem. The idealized model is not able to describe all the occurrences observed in real systems, but is able to describe those that are most essential. As soon as the validity of the idealized model is assessed, extensions and generalizations of the model are attempted in order to better describe the real system under consideration. Some extensions do not change the nature of the solutions obtained using the model, but others do.

The Black & Scholes model is one of the more successful idealized models currently in use. Since its introduction in 1973, a large amount of literature dealing with the extension of the Black & Scholes model has appeared. These extensions aim to relax assumptions that may not be realistic for real financial markets. Examples include

- option pricing with stochastic interest rate [4, 120];
- option pricing with a jump-diffusion/pure-jump stochastic process of stock price [13, 121];
- option pricing with a stochastic volatility [71, 72]; and
- option pricing with non-Gaussian distributions of log prices [7, 21] and with a truncated Lévy distribution [118].

We will briefly comment on general equations describing the time evolution of stock price and volatility [12] that is much more general than the Black & Scholes assumption of geometric Brownian motion. Our aim is to show how the complexity of equations increases when one or several of the Black & Scholes assumptions are relaxed. These general equations are

$$\frac{dY(t)}{Y(t)} = [r(t) - \lambda\mu_J]dt + \sigma(t)dW_Y(t) + J(t)dq(t) \qquad (15.4)$$

and

$$d\sigma^2(t) = [\theta_v - K_v\sigma^2(t)]dt + \sigma_v\sigma(t)dW_v(t), \qquad (15.5)$$

while the Black & Scholes assumption of geometric Brownian motion is, from (14.5),

$$\frac{dY(t)}{Y(t)} = \mu dt + \sigma dW(t); \qquad \sigma = \text{const.} \qquad (15.6)$$

Here $r(t)$ is the instantaneous spot interest rate, λ the frequency of jumps per year, $\sigma^2(t)$ the diffusion component of return variance, $W_Y(t)$ and $W_v(t)$ standard Wiener processes with covariance $\text{cov}[dW_Y(t), dW_v(t)] = \rho dt$, $J(t)$ the percentage jump size with unconditional mean μ_J, $q(t)$ a Poisson process with intensity λ, and K_v, θ_v and σ_v parameters of the diffusion component of return variance $\sigma^2(t)$.

It is worth pointing out that the increase in complexity is not only technical, but also conceptual. This is the case because the process is so general that it is no longer possible to build a simple replicating portfolio, or to perfectly hedge an 'optimal' portfolio. The elegance of the Black & Scholes solution is lost in real markets.

15.5 Summary

Complete knowledge of statistical properties of asset return dynamics is essential for fundamental and applied reasons. Such knowledge is crucial for the building and testing of a statistical model of a financial market. In spite of more than 50 years of effort, this goal has not yet been achieved.

The practical relevance of the resolution of the problem of the statistical properties of asset return dynamics is related to the optimal resolution of the rational pricing of an option. This is a financial activity that is extremely important in present-day financial markets. We saw that the dynamical properties of asset return dynamics – such as the continuous or discontinuous nature of its changes, the random character of its volatility, and the knowledge of the pdf function of asset returns – need to be known in order to adequately pose, and possibly solve, the option-pricing problem.

Statistical and theoretical physicists can contribute to the resolution of these scientific problems by sharing – with researchers in the other disciplines involved – the background in critical phenomena, disordered systems, scaling, and universality that has been developed over the last 30 years.

Appendix A: Notation guide

Chapter 1

x	income of a given individual
y	number of people having income x or greater
v	exponent of Pareto law
x_i	random variable
S_n	sum of n random variables
$P(x)$	probability density function of the random variable x
α	index of the Lévy stable distribution
d	dimension of a chaotic attractor
$\sim$	symbol to denote asymptotic equality

Chapter 2

t	time
Y_t	price of a financial asset at time t
$E\{x\}$	expected value of the variable x
$E\{x|y_1, y_2, y_3, \ldots\}$	expected value of x conditional on the occurrence of $y_1, y_2, y_3, \ldots$
$K^{(n)}$	bit length of the shortest computer program able to print a given string of length n

Chapter 3

x_i	random variable
n	number of random variables
S_n	sum of n random variables
$E\{f(x)\}$	average value of $f(x)$
δ_{ij}	Kronecker delta
Δt	time step

Chapter 3 (cont.)

$x(n\Delta t)$	sum of n random variables, each one occurring after a time step Δt
s^2	second moment of a dichotomic variable x_i
D	diffusion constant
$\otimes$	convolution symbol
σ_n	standard deviation of S_n
U_i	truncated random variable
ϵ	small number
$\tilde{x}$	scaled variable
$P(x)$	probability density function
$\tilde{P}(\tilde{x})$	scaled probability density function
π	pi
$F_n(S)$	distribution function of a scaled $\tilde{S}(n)$
$\Phi(S)$	distribution function of a Gaussian process
$Q_j(S)$	polynomial encountered in convergence studies
r_i	third moment of the absolute value of x_i
s_n^2	sum of n variances σ_i^2

Chapter 4

$P(x)$	probability density function
$\varphi(q)$	characteristic function
$\mathscr{F}[f(x)]$	operator indicating the Fourier transform of $f(x)$
$F(q)$	Fourier transform of $f(x)$
$\varphi_n(q)$	characteristic function of random variable S_n
$P_{\mathrm{L}}(x)$	symmetric Lévy stable distribution
γ	scale factor of the Lévy distribution
μ	average value of a random variable
β	asymmetry parameter of Lévy distribution
$\Gamma(x)$	Gamma function
$\tilde{S}_n$	scaled variable
$\tilde{P}(\tilde{S}_n)$	scaled probability density function
$\varphi_k(q)$	characteristic function of elementary random variable concurring to an infinitely divisible random variable
ℓ	length
k	integer number
$Z(t)$	price change at time t

Chapter 5

$Y(t)$	price of a financial asset at time t
$Z(t)$	price change at time t
$Z_D(t)$	deflated or discounted price change
$D(t)$	deflating or discounting time series
$R(t)$	return at time t
$S(t)$	successive differences of the natural logarithm of price

Chapters 6 and 7

$E\{f(x)\}$	expected value of $f(x)$
$f(x,t)$	probability density of observing x at time t
$f(x_1, x_2; t_1, t_2)$	joint probability density of observing x_1 at time t_1 and x_2 at time t_2
$f(x_1; t_1 \mid x_2; t_2)$	conditional probability density of observing x_2 at t_2 after observing x_1 at t_1
μ	average value of the random process
$R(t_1, t_2)$	autocorrelation function
$\tau \equiv t_2 - t_1$	time lag
$C(t_1, t_2)$	autocovariance
τ_c	characteristic time
τ_0	time scale
ν	exponent
η	exponent
τ^*	typical time
σ^2	variance
f	frequency
$S(f)$	power spectrum
$\sigma(t)$	standard deviation at different time horizons

Chapters 8 and 9

$Y(t)$	price of a financial asset at time t
$S(t)$	difference of the logarithm of price
x	random variable
y_i	random variables
Z	random variable
$P(Z)$	probability density function
n, k, m	integers
C_n	constant
$\Gamma(x)$	Gamma function
$\Omega(t)$	directing process

Chapters 8 and 9 (cont.)

$P_L(x)$	Lévy distribution
ℓ	truncation length
$P_G(x)$	Gaussian distribution
α	index of the Lévy distribution
γ	scale factor of the Lévy distribution
c	constant
$n_\times$	number of i.i.d. variables needed to observe a crossover between Lévy and Gaussian regimes
$\tilde{S}$	scaled variable
$\tilde{P}(\tilde{S})$	scaled probability density function
$\varphi(q)$	characteristic function
$Z(t)$	index changes
σ	standard deviation
g_i	normalized difference of the logarithm of price of company i
$F(g)$	cumulative distribution
σ_i	volatility (standard deviation) of company i

Chapter 10

x_t	discrete random variable
$f_t(x)$	conditional (on t) probability density function
t	time
σ_t^2	variance of the variable x_t at time t
p	integer
$\alpha_0, \alpha_1, \ldots, \alpha_q$	parameters
κ	kurtosis of the stochastic process
$P(x)$	asymptotic probability density function
$\beta_1, \beta_2, \ldots, \beta_p$	parameters
η_t	random variable
$R(\tau)$	autocorrelation function
τ	characteristic time scale
$M(t, k)$	polynomials of the random variables η_t
n	integer
$\text{cov}(x, y)$	covariance of random variables x and y
$\langle x \rangle$	average value of x
$S_t^{(m)}$	aggregate GARCH(1,1) innovation
$\alpha_0^{(m)}, \alpha_1^{(m)}$	parameter of the m aggregate GARCH(1,1) process
$\beta^{(m)}$	parameter of the m aggregate GARCH(1,1) process
Δt	time horizon

Chapter 11

L	characteristic length
V	characteristic flow velocity
v	kinematic viscosity
P	pressure
$\mathbf{r}$	position vector
$\Delta V(\ell)$	velocity increment
η	power spectrum exponent
$P_G(0)$	Gaussian probability of return to the origin
$P_{\Delta t}(V = 0)$	probability of return to the origin at time horizon Δt
ϵ	mean energy dissipation rate per unit mass
ℓ	length scale
$[x]$	physical dimension of the observable x
a, b	exponents

Chapters 12 and 13

S_i	differences of the logarithm of closure price of stock i
ρ_{ij}	correlation coefficient between stocks i and j
σ	standard deviation of ρ_{ij}
$\langle \rho_{ij} \rangle$	average value of ρ_{ij}
δ_{ij}	normalized position of ρ_{ij} with respect to average value
d_{ij}	Euclidean distance between stocks i and j
$\hat{d}_{ij}$	ultrametric distance between stocks i and j
R_n	one period return of the n asset
R_{n0}	risk-free and factor-risk premia mean return of the n asset
B	$n \times k$ matrix of factor weights
ξ_k	time series of the k factor
ϵ_n	specific risk of the n asset

Chapter 14

Y	price of a financial asset
K	strike price of an option
T	delivery (or maturity) date of an option
$C(Y, t)$	rational price of an option
Δ_h	number of shares for each option in a riskless portfolio
ϕ	value of a portfolio
μ	instantaneous expected return per unit time
σ^2	instantaneous variance per unit time
W	Wiener process
r	market interest rate

Chapter 14 (cont.)

$y(x, t')$ Black & Scholes transformation

d_1 and d_2 Black & Scholes variables

Chapter 15

σ_{imp} implied volatility

$r(t)$ instantaneous spot interest rate

λ frequency of jumps per year

$\sigma(t)$ diffusive part of the volatility

σ_v standard deviation of volatility fluctuations

K_v, Θ_v parameters of the diffusion component of return variance

Appendix B: Martingales

A new concept was introduced in probability theory about half a century ago – the martingale. J. Ville introduced the term, but its roots go back to P. Lévy in 1934 (see ref. [77]). The first complete theory of martingales was formulated by Doob [42].

Let the observed process be denoted by S_n. Let $\mathscr{F}_n$ represent a family of information sets (technically, a 'filtration'). Using a given set of information $\mathscr{F}$, one can generate a 'forecast' of the outcome S_n

$$E\{S_n | \mathscr{F}_{n-1}\}.$$

S_n is a martingale relative to $(\{\mathscr{F}_n\}, \mathscr{P})$ if

(i) S_n is known, given $\mathscr{F}_n$ (the technical term is that S_n is adapted),
(ii) $E\{|S_n|\} < \infty$, $\forall n$ (unconditional forecasts are finite), and
(iii) $E\{S_n | \mathscr{F}_{n-1}\} = S_{n-1}$, a.s. ($n \geq 1$) (i.e., the best forecast of unobserved future values is the last observation of S_{n-1}).

Here $\mathscr{P}$ is a probability measure and all expectations $E\{\bullet\}$ are assumed to be taken with respect to $\mathscr{P}$. A martingale is defined relative to a given filtration and probability measure. The essence of a martingale is to be a zero-drift stochastic process.

This concept is fundamental in mathematical finance because, e.g., in a world in which interest rates are zero and there are no arbitrage opportunities, there exists a unique equivalent martingale measure under which the price of any non-income-producing security equals its expected future price [70].

References

[1] V. Akgiray, 'Conditional Heteroskedasticity in Time Series of Stock Returns: Evidence and Forecasts', *J. Business* **62**, 55–80 (1989).

[2] L. A. N. Amaral, S. V. Buldyrev, S. Havlin, H. Leschhorn, P. Maass, M. A. Salinger, H. E. Stanley, and M. H. R. Stanley, 'Scaling Behavior in Economics: I. Empirical Results for Company Growth', *J. Phys. I France* **7**, 621–633 (1997).

[3] L. A. N. Amaral, S. V. Buldyrev, S. Havlin, M. A. Salinger, and H. E. Stanley, 'Power Law Scaling for a System of Interacting Units with Complex Internal Structure', *Phys. Rev. Lett.* **80**, 1385–1388 (1998).

[4] K. Amin and R. Jarrow, 'Pricing Options on Risky Assets in a Stochastic Interest Rate Economy', *Mathematical Finance* **2**, 217–237 (1992).

[5] P. W. Anderson, J. K. Arrow and D. Pines, eds., *The Economy as an Evolving Complex System* (Addison-Wesley, Redwood City, 1988).

[6] A. Arneodo, J. F. Muzy, and D. Sornette, '"Direct" Causal Cascade in the Stock Market', *Eur. Phys. J. B* **2**, 277–282 (1998).

[7] E. Aurell, R. Baviera, O. Hammarlid, M. Serva, and A. Vulpiani, 'A General Methodology to Price and Hedge Derivatives in Incomplete Markets', *Int. J. Theor. Appl. Finance* (in press).

[8] L. Bachelier, 'Théorie de la spéculation' [Ph.D. thesis in mathematics], *Annales Scientifiques de l'Ecole Normale Supérieure* **III-17**, 21–86 (1900).

[9] R. T. Baillie and T. Bollerslev, 'Conditional Forecast Densities from Dynamics Models with GARCH Innovations', *J. Econometrics* **52**, 91–113 (1992).

[10] P. Bak, K. Chen, J. Scheinkman, and M. Woodford, 'Aggregate Fluctuations from Independent Sectoral Shocks: Self-Organized Criticality in a Model of Production and Inventory Dynamics', *Ricerche Economiche* **47**, 3–30 (1993).

[11] P. Bak, M. Paczuski, and M. Shubik, 'Price Variations in a Stock Market with Many Agents', *Physica A* **246**, 430–453 (1997).

[12] G. Bakshi, C. Cao, and Z. Chen, 'Empirical Performance of Alternative Option Pricing Models', *J. Finance* **52**, 2003–2049 (1997).

[13] D. Bates, 'The Crash of 87: Was it Expected? The Evidence from Options Markets', *J. Finance* **46**, 1009–1044 (1991).

[14] R. Baviera, M. Pasquini, M. Serva, and A. Vulpiani, 'Optimal Strategies for Prudent Investors', *Int. J. Theor. Appl. Finance* **1**, 473–486 (1998).

[15] J. P. Benzécri, *L'analyse des données 1, La Taxinomie* (Dunod, Paris, 1984).

137

[16] H. Bergström, 'On Some Expansions of Stable Distributions', *Ark. Mathematicae II* **18**, 375–378 (1952).

[17] A. C. Berry, 'The Accuracy of the Gaussian Approximation to the Sum of Independent Variates', *Trans. Amer. Math. Soc.* **49**, 122–136 (1941).

[18] F. Black and M. Scholes, 'The Pricing of Options and Corporate Liabilities', *J. Polit. Econ.* **81**, 637–654 (1973).

[19] R. C. Blattberg and N. J. Gonedes, 'A Comparison of the Stable and Student Distributions as Statistical Model for Stock Prices', *J. Business* **47**, 244–280 (1974).

[20] T. Bollerslev, 'Generalized Autoregressive Conditional Heteroskedasticity', *J. Econometrics* **31**, 307–327 (1986).

[21] J.-P. Bouchaud and D. Sornette, 'The Black & Scholes Option Pricing Problem in Mathematical Finance: Generalization and Extensions for a Large Class of Stochastic Processes', *J. Phys. I France* **4**, 863–881 (1994).

[22] J.-P. Bouchaud and M. Potters, *Théories des Risques Financiers* (Eyrolles, Aléa-Saclay, 1997).

[23] J.-P. Bouchaud and R. Cont, 'A Langevin Approach to Stock Market Fluctuations and Crashes', *Eur. Phys. J. B* **6**, 543–550 (1998).

[24] S. J. Brown, 'The Number of Factors in Security Returns', *J. Finance* **44**, 1247–1262 (1989).

[25] G. Caldarelli, M. Marsili, and Y.-C. Zhang, 'A Prototype Model of Stock Exchange', *Europhys. Lett.* **40**, 479–483 (1997).

[26] J. Y. Campbell, A. W. Lo, and A. C. MacKinlay, *The Econometrics of Financial Markets* (Princeton University Press, Princeton, 1997).

[27] M. Cassandro and G. Jona-Lasinio, 'Critical Point Behaviour and Probability Theory', *Adv. Phys.* **27**, 913–941 (1978).

[28] G. J. Chaitin, 'On the Length of Programs for Computing Finite Binary Sequences', *J. Assoc. Comp. Math.* **13**, 547–569 (1966).

[29] D. Challet and Y. C. Zhang, 'On the Minority Game: Analytical and Numerical Studies', *Physica A* **256**, 514–532 (1998).

[30] P. L. Chebyshev, 'Sur deux théorèmes relatifs aux probabilités', *Acta Math.* **14**, 305–315 (1890).

[31] P. Cizeau, Y. Liu, M. Meyer, C.-K. Peng, and H. E. Stanley, 'Volatility Distribution in the S&P 500 Stock Index', *Physica A* **245**, 441–445 (1997).

[32] P. K. Clark, 'A Subordinated Stochastic Process Model with Finite Variance for Speculative Prices', *Econometrica* **41**, 135–256 (1973).

[33] E. R. Cohen and B. N. Taylor, 'The 1986 Adjustment of the Fundamental Physical Constants', *Rev. Mod. Phys.* **59**, 1121–1148 (1987).

[34] G. Connor and R. A. Korajczyk, 'A Test for the Number of Factors in an Approximate Factor Model', *J. Finance* **48**, 1263–1291 (1993).

[35] R. Cont, M. Potters, and J.-P. Bouchaud, 'Scaling in Stock Market Data: Stable Laws and Beyond', in *Scale Invariance and Beyond*, edited by B. Dubrulle, F. Graner, and D. Sornette (Springer, Berlin, 1997).

[36] R. Cont, 'Scaling and Correlation in Financial Data', Cond.-Mat. preprint server 9705075.

[37] J. C. Cox, S. A. Ross, and M. Rubinstein, 'Option Pricing: A Simplified Approach', *J. Financial Econ.* **7**, 229–263 (1979).

[38] P. H. Cootner, ed., *The Random Character of Stock Market Prices* (MIT Press, Cambridge MA, 1964).

[39] A. Crisanti, G. Paladin, and A. Vulpiani, *Products of Random Matrices in Statistical Physics* (Springer-Verlag, Berlin, 1993).

[40] L. Crovini and T. J. Quinn, eds., *Metrology at the Frontiers of Physics and Technology* (North-Holland, Amsterdam, 1992).

[41] M. M. Dacorogna, U. A. Müller, R. J. Nagler, R. B. Olsen, and O. V. Pictet, 'A Geographical Model for the Daily and Weekly Seasonal Volatility in the Foreign Exchange Market', *J. Intl Money and Finance* **12**, 413–438 (1993).

[42] J. L. Doob, *Stochastic Processes* (J. Wiley & Sons, New York, 1953).

[43] F. C. Drost and T. E. Nijman, 'Temporal Aggregation of GARCH Processes', *Econometrica* **61**, 909–927 (1993).

[44] D. Duffie and C. Huang, 'Implementing Arrow-Debreu Equilibria by Continuous Trading of a Few Long-Lived Securities', *Econometrica* **53**, 1337–1356 (1985).

[45] D. Duffie, *Dynamic Asset Pricing Theory, Second edition* (Princeton University Press, Princeton, 1996).

[46] P. Dutta and P. M. Horn, 'Low-Frequency Fluctuations in Solids: $1/f$ Noise', *Rev. Mod. Phys.* **53**, 497–516 (1981).

[47] E. Eberlein and U. Keller, 'Hyperbolic Distributions in Finance', *Bernoulli* **1**, 281–299 (1995).

[48] A. Einstein, 'On the Movement of Small Particles Suspended in a Stationary Liquid Demanded by the Molecular-Kinetic Theory of Heat', *Ann. Physik* **17**, 549–560 (1905).

[49] E. J. Elton and M. J. Gruber, *Modern Portfolio Theory and Investment Analysis* (J. Wiley and Sons, New York, 1995).

[50] R. F. Engle, 'Autoregressive Conditional Heteroskedasticity with Estimates of the Variance of U.K. Inflation', *Econometrica* **50**, 987–1002 (1982).

[51] C. G. Esséen, 'Fourier Analysis of Distributions Functions. A Mathematical Study of the Laplace–Gaussian Law', *Acta Math.* **77**, 1–125 (1945).

[52] E. F. Fama, 'The Behavior of Stock Market Prices', *J. Business* **38**, 34–105 (1965).

[53] E. F. Fama, 'Efficient Capital Markets: A Review of Theory and Empirical Work', *J. Finance* **25**, 383–417 (1970).

[54] E. F. Fama, 'Efficient Capital Markets: II', *J. Finance* **46**, 1575–1617 (1991).

[55] J. D. Farmer, 'Market Force, Ecology, and Evolution', Adap-Org preprint server 9812005.

[56] W. Feller, *An Introduction to Probability Theory and Its Applications, Vol. 1, Third edition* (J. Wiley & Sons, New York, 1968).

[57] W. Feller, *An Introduction to Probability Theory and Its Applications, Vol. 2, Second edition* (J. Wiley & Sons, New York, 1971).

[58] S. Figlewski, 'Options Arbitrage in Imperfect Markets', *J. Finance* **64**, 1289–1311 (1989).

[59] M. E. Fisher, 'The Theory of Critical Point Singularities', in *Proc. Enrico Fermi School on Critical Phenomena*, edited by M. S. Green (Academic Press, London and New York, 1971), pp. 1–99.

[60] K. R. French, 'Stock Returns and the Weekend Effect', *J. Financial Econ.* **8**, 55–69 (1980).

[61] U. Frisch, *Turbulence: The Legacy of A. N. Kolmogorov* (Cambridge University Press, Cambridge, 1995).

[62] S. Galluccio and Y. C. Zhang, 'Products of Random Matrices and Investment Strategies', *Phys. Rev. E* **54**, R4516–R4519 (1996).

[63] S. Galluccio, J.-P. Bouchaud, and M. Potters, 'Rational Decisions, Random Matrices and Spin Glasses', *Physica A* **259**, 449–456 (1998).

[64] S. Ghashghaie, W. Breymann, J. Peinke, P. Talkner, and Y. Dodge, 'Turbulent Cascades in Foreign Exchange Markets', *Nature* **381**, 767–770 (1996).

[65] B. V. Gnedenko, 'On the Theory of Domains of Attraction of Stable Laws', *Uchenye Zapiski Moskov. Gos. Univ. Matematkia* **45**, 61–72 (1940).

[66] B. V. Gnedenko and A. N. Kolmogorov, *Limit Distributions for Sums of Independent Random Variables* (Addison-Wesley, Cambridge MA, 1954).

[67] P. Gopikrishnan, M. Meyer, L. A. N. Amaral, and H. E. Stanley, 'Inverse Cubic Law for the Distribution of Stock Price Variations', *Eur. Phys. J. B* **3**, 139–140 (1998).

[68] P. Gopikrishnan, M. Meyer, L. A. N. Amaral, V. Plerou, and H. E. Stanley, 'Scaling and Volatility Correlations in the Stock Market', Cond.-Mat. preprint server 9905305; *Phys. Rev. E* (in press).

[69] T. Guhr, A. Müller-Groeling, and H. A. Weidenmüller, 'Random-Matrix Theories in Quantum Physics: Common Concepts', *Phys. Reports* **299**, 189–425 (1998).

[70] J. M. Harrison and D. M. Kreps, 'Martingales and Arbitrage in Multiperiod Securities Markets', *J. Econ. Theor.* **20**, 381–408 (1979).

[71] S. Heston, 'A Closed-form Solution for Options with Stochastic Volatility with Application to Bond and Currency Options', *Rev. Financial Studies* **6**, 327–343 (1993).

[72] J. Hull and A. White, 'The Pricing of Options with Stochastic Volatilities', *J. Finance* **42**, 281–300 (1987).

[73] J. C. Hull, *Options, Futures, and Other Derivatives, Third edition* (Prentice-Hall, Upper Saddle River NJ, 1997).

[74] J. E. Ingersoll, Jr, *Theory of Financial Decision Making* (Rowman & Littlefield, Savage MD, 1987).

[75] K. Ito, 'On Stochastic Differential Equations', *Mem. Amer. Math. Soc.* **4**, 1–51 (1951).

[76] L. P. Kadanoff, 'From Simulation Model to Public Policy: An examination of Forrester's Urban Dynamics', *Simulation* **16**, 261–268 (1971).

[77] I. Karatzas and S. E. Shreve, *Brownian Motion and Stochastic Calculus* (Springer-Verlag, Berlin, 1988).

[78] J. Kertész and I. Kondor, eds., *Econophysics: Proc. of the Budapest Workshop* (Kluwer Academic Press, Dordrecht, 1999).

[79] M. S. Keshner, '1/f Noise', *Proc. IEEE* **70**, 212–218 (1982).

[80] A. Ya. Khintchine and P. Lévy, 'Sur les loi stables', *C. R. Acad. Sci. Paris* **202**, 374–376 (1936).

[81] A. Ya. Khintchine, 'Zur Theorie der unbeschränkt teilbaren Verteilungsgesetze', *Rec. Math. [Mat. Sbornik] N. S.* **2**, 79–120 (1937).

[82] A. N. Kolmogorov, 'The Local Structure of Turbulence in Incompressible Viscous Fluid for Very Large Reynolds Number', *Dokl. Akad. Nauk. SSSR* **30**, 9–13 (1941) [reprinted in *Proc. R. Soc. Lond. A* **434**, 9–13 (1991)].

[83] A. N. Kolmogorov, 'On Degeneration of Isotropic Turbulence in an Incompressible Viscous Liquid', *Dokl. Akad. Nauk. SSSR* **31**, 9538–9540 (1941).

[84] A. N. Kolmogorov, 'Dissipation of Energy in Locally Isotropic Turbulence',

Dokl. Akad. Nauk. SSSR **32**, 16–18 (1941) [reprinted in *Proc. R. Soc. Lond. A* **434** 15–17 (1991)].

[85] A. N. Kolmogorov, 'Three Approaches to the Quantitative Definition of Information', *Problems of Information Transmission* **1**, 4 (1965).

[86] I. Koponen, 'Analytic Approach to the Problem of Convergence of Truncated Lévy Flights towards the Gaussian Stochastic Process', *Phys. Rev. E* **52**, 1197–1199 (1995).

[87] L. Laloux, P. Cizeau, J.-P. Bouchaud, and M. Potters, 'Noise Dressing of Financial Correlation Matrices', *Phys. Rev. Lett.* **83**, 1467–1470 (1999).

[88] K. Lauritsen, P. Alström, and J.-P. Bouchaud, eds. Application of Physics in Financial Analysis, *Int. J. Theor. Appl. Finance* [special issue] (in press).

[89] Y. Lee, L. A. N. Amaral, D. Canning, M. Meyer, and H. E. Stanley, 'Universal Features in the Growth Dynamics of Complex Organizations', *Phys. Rev. Lett.* **81**, 3275–3278 (1998).

[90] M. Lévy, H. Lévy, and S. Solomon, 'Microscopic Simulation of the Stock-Market – The Effect of Microscopic Diversity', *J. Phys. I* **5**, 1087–1107 (1995).

[91] M. Lévy and S. Solomon, 'Power Laws Are Logarithmic Boltzmann Laws', *Intl J. Mod. Phys. C* **7**, 595–601 (1996).

[92] P. Lévy, *Calcul des probabilités* (Gauthier-Villars, Paris, 1925).

[93] W. Li, 'Absence of $1/f$ Spectra in Dow Jones Daily Average', *Intl J. Bifurcations and Chaos* **1**, 583–597 (1991).

[94] J. W. Lindeberg, 'Eine neue Herleitung des Exponentialgesetzes in der Wahrscheinlichkeitsrechnung', *Mathematische Zeitschrift* **15**, 211–225 (1922).

[95] Y. Liu, P. Cizeau, M. Meyer, C.-K. Peng, and H. E. Stanley, 'Quantification of Correlations in Economic Time Series', *Physica A* **245**, 437–440 (1997).

[96] Y. Liu, P. Gopikrishnan, P. Cizeau, M. Meyer, C.-K. Peng, and H. E. Stanley, 'The Statistical Properties of the Volatility of Price Fluctuations', *Phys. Rev. E* **59**, 1390–1400 (1999).

[97] A. W. Lo and A. C. Mackinlay, 'When Are Contrarian Profits Due to Stock Market Overreaction?' *Rev. Financial Stud.* **3**, 175–205 (1990).

[98] T. Lux, 'Time Variation of Second Moments from a Noise Trader Infection Model', *J. Econ. Dyn. Control* **22**, 1–38 (1997).

[99] T. Lux, 'The Socio-Economic Dynamics of Speculative Markets: Interacting Agents, Chaos, and the Fat Tails of Return Distributions', *J. Econ. Behav. Organ* **33**, 143–165 (1998).

[100] T. Lux and M. Marchesi, 'Scaling and Criticality in a Stochastic Multi-Agent Model of a Financial Market', *Nature* **397**, 498–500 (1999).

[101] E. Majorana, 'Il valore delle leggi statistiche nella fisica e nelle scienze sociali', *Scientia* **36**, 58–66 (1942).

[102] B. B. Mandelbrot, 'The Variation of Certain Speculative Prices', *J. Business* **36**, 394–419 (1963).

[103] B. B. Mandelbrot, *The Fractal Geometry of Nature* (W. H. Freeman, San Francisco, 1982).

[104] B. B. Mandelbrot, *Fractals and Scaling in Finance* (Springer-Verlag, New York, 1997).

[105] R. N. Mantegna, 'Lévy Walks and Enhanced Diffusion in Milan Stock Exchange', *Physica A* **179**, 232–242 (1991).

[106] R. N. Mantegna, 'Fast, Accurate Algorithm for Numerical Simulation of Lévy Stable Stochastic Processes', *Phys. Rev. E* **49**, 4677–4683 (1994).

[107] R. N. Mantegna, 'Degree of Correlation Inside a Financial Market', in *Applied Nonlinear Dynamics and Stochastic Systems near the Millennium,* edited by J. B. Kadtke and A. Bulsara (AIP Press, New York, 1997), pp. 197–202.

[108] R. N. Mantegna, 'Hierarchical Structure in Financial Markets', Cond.-Mat. preprint server 9802256; *Eur. Phys. J. B* **11**, 193–197 (1999).

[109] R. N. Mantegna, ed., *Proceedings of the International Workshop on Econophysics and Statistical Finance, Physica A* [special issue] **269**, (1999).

[110] R. N. Mantegna and H. E. Stanley, 'Stochastic Process with Ultraslow Convergence to a Gaussian: the Truncated Lévy Flight', *Phys. Rev. Lett.* **73**, 2946–2949 (1994).

[111] R. N. Mantegna and H. E. Stanley, 'Scaling Behavior in the Dynamics of an Economic Index', *Nature* **376**, 46–49 (1995).

[112] R. N. Mantegna and H. E. Stanley, 'Turbulence and Financial Markets', *Nature* **383**, 587–588 (1996).

[113] R. N. Mantegna and H. E. Stanley, 'Stock Market Dynamics and Turbulence: Parallel Analysis of Fluctuation Phenomena', *Physica A* **239**, 255–266 (1997).

[114] R. N. Mantegna and H. E. Stanley, 'Physics Investigation of Financial Markets', in *Proceedings of the International School of Physics 'Enrico Fermi', Course CXXXIV*, edited by F. Mallamace and H. E. Stanley (IOS Press, Amsterdam, 1997).

[115] H. Markowitz, *Portfolio Selection: Efficient Diversification of Investment* (J. Wiley, New York, 1959).

[116] M. Marsili, S. Maslov, and Y.-C. Zhang, 'Dynamical Optimization Theory of a Diversified Portfolio', *Physica A* **253**, 403–418 (1998).

[117] S. Maslov and Y.-C. Zhang, 'Probability Distribution of Drawdowns in Risky Investments', *Physica* **262**, 232–241 (1999).

[118] A. Matacz, 'Financial Modeling on Option Theory with the Truncated Lévy Process', Working Paper, School of Mathematics and Statistics, University of Sydney, Report 97-28 (1997).

[119] M. Mehta, *Random Matrices* (Academic Press, New York, 1995).

[120] R. C. Merton, 'Theory of Rational Option Pricing', *Bell J. Econ. Management Sci.* **4**, 141–183 (1973).

[121] R. C. Merton, 'Option Pricing When Underlying Stock Returns Are Discontinuous', *J. Financial Econ.* **3**, 125–144 (1976).

[122] R. C. Merton, *Continuous-Time Finance* (Blackwell, Cambridge MA, 1990).

[123] M. Mézard, G. Parisi, and M. A. Virasoro, *Spin Glass Theory and Beyond* (World Scientific, Singapore, 1987).

[124] A. S. Monin and A. M. Yaglom, *Statistical Fluid Mechanics: Mechanics of Turbulence, Vol. 1* (The MIT Press, Cambridge, 1971).

[125] E. W. Montroll and W. W. Badger, *Introduction to Quantitative Aspects of Social Phenomena* (Gordon and Breach, New York, 1974).

[126] U. A. Müller, M. M. Dacorogna, R. B. Olsen, O. V. Pictet, M. Schwarz, and C. Morgenegg, 'Statistical Study of Foreign Exchange Rates, Empirical Evidence of a Price Change Scaling Law and Intraday Analysis', *J. Banking and Finance* **14**, 1189–1208 (1995).

[127] M. Musiela and M. Rutkowski, *Martingale Methods in Financial Modeling* (Springer, Berlin, 1997).

[128] M. M. Navier, 'Mémoire sur les Lois du Mouvement des Fluides', *Mém. Acad. Roy. Sci.* **6**, 389–440 (1923).

[129] A. Pagan, 'The Econometrics of Financial Markets', *J. Empirical Finance* **3**, 15–102 (1996).

[130] C. H. Papadimitriou and K. Steigitz, *Combinatorial Optimization* (Prentice-Hall, Englewood Cliffs NJ, 1982).

[131] A. Papoulis, *Probability, Random Variables, and Stochastic Processes, second edition* (McGraw-Hill, New York, 1984).

[132] V. Pareto, *Cours d'Economie Politique* (Lausanne and Paris, 1897).

[133] M. Pasquini and M. Serva, 'Clustering of Volatility as a Multiscale Phenomenon', Cond.-Mat. preprint server 9903334.

[134] V. Plerou, P. Gopikrishnan, B. Rosenow, L. A. N. Amaral, and H. E. Stanley, 'Universal and Non-Universal Properties of Cross-Correlations in Financial Time Series', *Phys. Rev. Lett.* **83**, 1471–1474 (1999).

[135] M. Potters, R. Cont, and J.-P. Bouchaud, 'Financial Markets as Adaptive Ecosystems', *Europhys. Lett.* **41**, 239–242 (1998).

[136] W. H. Press, 'Flicker Noise in Astronomy and Elsewhere', *Comments Astrophys.* **7**, 103–119 (1978).

[137] M. Raberto, E. Scalas, G. Cuniberti, and M. Riani, 'Volatility in the Italian Stock Market: An Empirical Study', *Physica A* **269**, 148–155 (1999).

[138] R. Rammal, G. Toulouse, and M. A. Virasoro, 'Ultrametricity for Physicists', *Rev. Mod. Phys.* **58**, 765–788 (1986).

[139] S. Ross, 'The Arbitrage Theory of Capital Asset Pricing', *J. Econ. Theory* **13**, 341–360 (1976).

[140] G. Samorodnitsky and M. S. Taqqu, *Stable Non-Gaussian Random Processes: Stochastic Models with Infinite Variance* (Chapman and Hall, New York, 1994).

[141] P. A. Samuelson, 'Proof that Properly Anticipated Prices Fluctuate Randomly', *Industrial Management Rev.* **6**, 41–45 (1965).

[142] A. H. Sato and H. Takayasu, 'Dynamic Numerical Models of Stock Market Price: From Microscopic Determinism to Macroscopic Randomness', *Physica A* **250**, 231–252 (1998).

[143] G. W. Schwert, 'Why Does Stock Market Volatility Change over Time?', *J. Finance* **44**, 1115–1153 (1989).

[144] M. F. Shlesinger, 'Comment on "Stochastic Process with Ultraslow Convergence to a Gaussian: The Truncated Lévy Flight",' *Phys. Rev. Lett.* **74**, 4959 (1995).

[145] D. Sornette, 'Large Deviations and Portfolio Optimization', *Physica A* **256**, 251–283 (1998).

[146] D. Sornette and A. Johansen, 'A Hierarchical Model of Financial Crashes', *Physica A* **261**, 581–598 (1998).

[147] H. E. Stanley, *Introduction to Phase Transitions and Critical Phenomena* (Oxford University Press, Oxford, 1971).

[148] M. H. R. Stanley, L. A. N. Amaral, S. V. Buldyrev, S. Havlin, H. Leschhorn, P. Maass, M. A. Salinger, and H. E. Stanley, 'Scaling Behavior in the Growth of Companies', *Nature* **379**, 804–806 (1996).

[149] D. Stauffer, 'Can Percolation Theory be Applied to the Stock Market?', *Ann. Phys.-Berlin* **7**, 529–538 (1998).

[150] D. Stauffer and T. J. P. Penna, 'Crossover in the Cont-Bouchaud Percolation Model for Market Fluctuations', *Physica A* **256**, 284–290 (1998).

[151] H. Takayasu, H. Miura, T. Hirabayashi, and K. Hamada, 'Statistical Properties of Deterministic Threshold Elements – The Case of Market Price', *Physica A* **184**, 127–134 (1992).

[152] H. Takayasu, A. H. Sato, and M. Takayasu, 'Stable Infinite Variance Fluctuations in Randomly Amplified Langevin Systems', *Phys. Rev. Lett.* **79**, 966–969 (1997).

[153] H. Takayasu and K. Okuyama, 'Country Dependence on Company Size Distributions and a Numerical Model Based on Competition and Cooperation', *Fractals* **6**, 67–79 (1998).

[154] C. Trzcinka, 'On the Number of Factors in the Arbitrage Pricing Model', *J. Finance* **41**, 347–368 (1986).

[155] N. Vandewalle and M. Ausloos, 'Coherent and Random Sequences in Financial Fluctuations', *Physica A* **246**, 454–459 (1997).

[156] M. B. Weissman, '$1/f$ Noise and Other Slow, Nonexponential Kinetics in Condensed Matter', *Rev. Mod. Phys.* **60**, 537–571 (1988).

[157] D. B. West, *Introduction to Graph Theory* (Prentice-Hall, Englewood Cliffs NJ, 1996).

[158] N. Wiener, 'Differential Space', *J. Math. Phys.* **2**, 131–174 (1923).

Index

145